Application of Solar Energy

Application of Solar Energy

Editor

Sabbas Radley

scitus academics

Application of Solar Energy
Edited by **Sabbas Radley**

ISBN: 978-1-68117-215-6
Library of Congress Control Number: 2016934758

SCITUS Academics LLC,
www.scitusacademics.com
Box No. 4766, 616 Corporate Way,
Suite 2, Valley Cottage,
NY 10989

Printed in United States of America on Acid Free Paper

Preface

Sunlight and heat are radiant energy from the sun using a variety of ever-changing technologies such as solar heating, photovoltaics, solar thermal energy, solar architecture and artificial photosynthesis. This energy moves from the sun and reaches the earth where people use solar cells to collect it and convert it to any desirable form of energy. Nonrenewable energy sources are used to boil water and form streams in conventional power plants so that turbines can rotate and water can generate electricity. The use of solar energy is also increasing in houses. Residential equipment can easily use solar-generated electricity. In addition to this solar power, solar heaters are used to supply hot water in the home. You can find PV glass modules or any other type of solar panel on the roofs of different buildings. These panels are used to supply different offices and other parts of the building with reliable electricity. Solar energy is used for ventilation purposes in many places. Not only does solar power improve the ventilation system of your home, but it can also help circulate water in any building. It will increase energy security by relying on an indigenous resource that is inexhaustible and mostly import-independent resource, increase sustainability, reduce pollution, reduce global warming costs and reduce fossil fuel prices.

This book is intended to focus on various aspects, such as household, commercial and industrial uses, of solar energy.

Table of Contents

1

CHAPTER

ANALYSIS OF THE POSSIBLE USE OF SOLAR PHOTOVOLTAIC ENERGY IN URBAN WATER SUPPLY SYSTEMS

Bojan Đurin [1,*] and Jure Margeta [2]

[1] *Faculty of Geotechnical Engineering, University of Zagreb, 42000 Varaždin, Hallerova Aleja 7, Croatia*
[2] *Faculty of Civil Engineering, Architecture & Geodesy, University of Split, 21000 Split, Matice Hrvatske 15, Croatia*

ABSTRACT

Because of the importance of water supply for the sustainability of urban areas, and due to the significant consumption of energy with prices increasing every day, an alternative solution for sustainable energy supply should be sought in the field of Renewable Energy Sources (RES). An innovative solution as presented in this paper has until now not been comprehensively analyzed. This work presents the solution with the application of a (Photovoltaic) PV generator. The main technological features, in addition to the designing methodology and case study are presented in this paper. The critical period approach has been used for the first time for system sizing. The application of this sizing method provides a high reliability of the proposed system. The obtained results confirm the assumption that the PV generator is a promising energy sustainable solution for urban water supply systems. The service reservoir, which acts as water and energy storage for the proposed system, provides the basis for a sustainable solution of water and energy supply. In accordance with the proposed, the reliability of such system is high. This concept of energy supply operation does not generate any atmospheric emission of greenhouse gases, which contributes significantly to the reduction of the impacts of climate changes. The proposed solution and designing methodology are widely applicable and in accordance with the characteristics of the water supply system and climate.

Keywords: urban water supply; solar energy; photovoltaic pumping; sustainable water supply; life cycle cost

1. URBAN WATER SYSTEMS AND SOLAR ENERGY

Urban water supply systems (UWSS) are defined as natural, modified and constructed elements of the urban water cycle, related to water supply, that can be found in towns. UWSS provide water to support human life, hygiene, health, safety, recreation and amenities. Natural elements of a system include local and regional water resources, while the modified and constructed elements include water intake, water supply pipes, pump stations, service reservoirs, and water treatment plants. The constructed water supply system is an integral part of the broader urban infrastructure system. Figure 1 shows the typical water supply management process, indicating location and transfer elements.

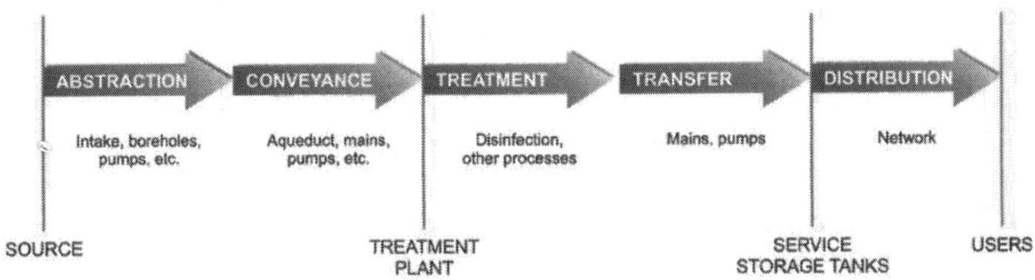

Figure 1. Typical elements of the water supply management system [1].

The biggest challenge for modern water utilities is to provide the required amount of water at affordable rates. UWSS activities are energy intensive and in general grow as the city and total water consumption are also expanded. Today, and especially in the future, UWSS management should contribute to the objective of sustainable urban development. A sustainable city minimizes energy use from traditional energy sources and reduces CO_2 emissions. Strategies to achieve sustainability include: (i) applying energy and water efficiency measures; (ii) shifting energy consumption as much as possible to off-peak periods; and (iii) the use of renewable energy. This paper will analyze the characteristics of application of PV technology for water pumping in the UWSS, or more specifically for bulk water supply. By using green energy, such as PV electric energy, cities minimize energy use from traditional electric energy sources and reduce CO_2 emission.

However, solar energy input is free. Broader objectives of the problem should be taken into consideration, as well as economical aspects of the use of green energy, instead of traditional energy from fossil fuels. This means that a system should be selected that will minimize the negative economical difference between the use of a green energy system and those of a traditional energy system by taking into consideration other benefits, since the supply of green energy is still more expensive than traditional energy. These other issues (benefits) which have to be taken into consideration are: energy-related environmental impacts, excessive dependence on specific energy forms, growing energy demand and supply problems, social cost independent of the cost of oil and other fossil fuels, locally available energy sources independent of regional energy network, reduction of CO_2 emission, *etc.*

Water is used in various amounts throughout the whole day, while solar irradiation occurs only during the sunlight periods of the day, (Figure 2) [2,3,4]. The duration of daily insolation varies throughout the year according to the climatic characteristics of the area being analyzed. Therefore, PV electric energy must be used when produced or Electric Energy Storage (EES) must be used, which will provide the supply of electric energy during the night, as well as during low solar radiation (cloudy period). In this way EES has a significant role in the implementation of PV electric energy supply to UWSS. Numerous technologies of energy storage are known today (batteries, flywheel, pressure vessels, *etc.*), which differ in: size, energy storage costs, efficiency, lifetime, costs per cycle, *etc.* [5,6]. For the time being, these EES are very expensive facilities and represent a solution for smaller consumers, but not for bigger facilities such as the UWSS.

The Main Pump Station (MPS) that pumps water into the clean water reservoir is generally a big energy consumer. Due to the pumping systems, energy consumption represents the biggest part of energy expenses in the water sector—sometimes up to 90% [7]. Significant use of energy is required, since all water has to be brought to a higher elevation in order to produce the required pressures to water users in the water supply area. This paper analyzes whether it is possible to use PV technology for urban water supply, whether water service reservoir can be used as EES and how the system balancing period affects its features.

The use of solar energy for water pumping has already been well addressed in literature, for example [8,9,10]. However, at present, there are no significant publications on a more thorough consideration of possible applications of PV generators in UWSS.

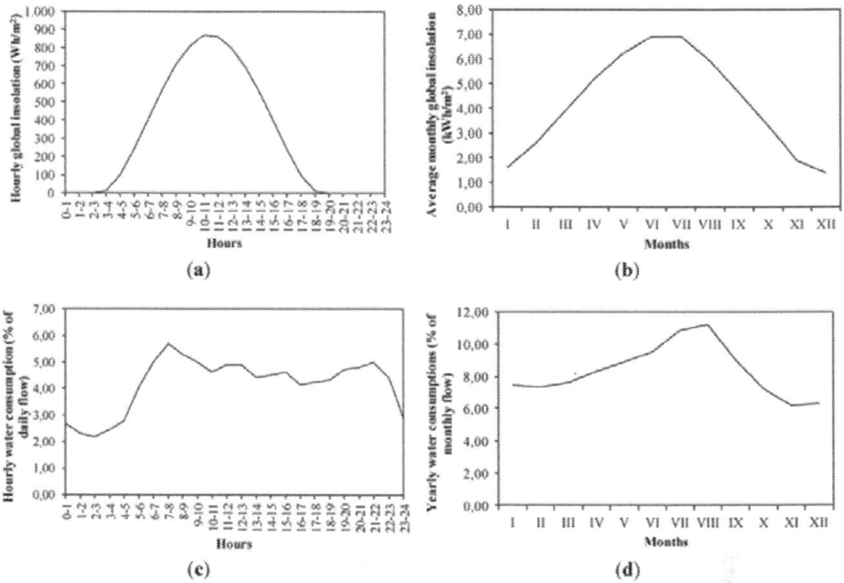

Figure 2. Examples of solar energy irradiation and water demand pattern in the urban areas. (**a**) Varaždin, Croatia, 15 July 2005; (**b**) Šibenik, Croatia, 1961–1980; (**c**) Varaždin, Croatia, 12 July 2013; (**d**) Šibenik, Croatia, 2013.

This particularly applies to the analysis of the impact of balancing period on the solution characteristics. PV energy can be used as an additional source of "green energy," or as the only source of electric energy required for pumping water. This paper analyzes the application of the PV generator as the sole source of energy to operate the MPS.

2. DESIGN METHODOLOGY

2.1. Critical Period Approach

Service reservoir and associated pump station, which deliver water into the reservoir when energy from the energy power system is used, are usually designed according to deterministic guidelines that specify minimal water service reservoir and optimal pump system capacity. The required capacity of the reservoir and pumping system is determined according to daily consumer demand for water in design period $V_{WS(i)}$ and hourly demand $Q_{WS(i)}$ in the day of maximum demand, the *critical demand period*. Design period t_b is usually the period with the highest consumption of water during the planning year. However, the design period can be longer: two or more days, up to a maximum of one week.

The size of the reservoir depends on the regime of water consumption in a settlement, and on the planned work of the pumping station. The work of pumping station, supplied with electric energy from the regional power system, can be continuous for 24 hours or more intense during the night when electricity is cheaper. Various combinations are possible that seek to optimize the operation of the pumping system, its capacity Q_{PS} and reservoir capacity V_{op}. The capacity of such a reservoir increases for fire and emergency storage, in accordance with the respective norms. Such guidelines have to accommodate a large range of possible conditions, meaning that systems are potentially overdesigned, but reliable.

In the case when PV electric energy is used, the design is more complex, because the available energy for pump system operation is variable throughout the year and during the day, depending on daily insolation. In this case, the designer must determine the PV generator power $P_{el,PV}$ that will provide enough electricity for re-pumping water into the reservoir during the entire planning period (planning year), according to daily water consumption in a settlement $V_{WS(i)}$. In addition, the duration of sunshine $T_{S(i)}$ determines the period of operation of the pumping station. In this case, the critical energy supply period, or the period for calculation of the PV generator power, is one in which the relationship between hydraulic energy required for re-pumping water $E_{H(i)}$ and the available solar radiation $E_{S(i)}$ is the smallest.

Consequently, the critical period for calculating the capacity of the pumping station $Q_{PS(i)}$ is the one in which the relationship between the daily consumption of water in a settlement $V_{WS(i)}$ and the period of active insolation $T_{S(i)}$ is the greatest.

If the design period t_b is longer than one day, the impact of one extreme day of low insolation (usually due to cloudiness) on the capacity of the PV generator decreases, so that lower power of the PV generator is required and thereby lower costs. With a longer design period, the impact of extremely low insolation on calculation results will be smaller, so the solution will be safer and more rational. In this case they are longer critical periods.

In this paper we used the approach based on *critical design period*. This approach includes design elements of the solution: PV generator, pump station and service reservoir based on the critical period of operation of each one. It is a deterministic approach which accommodates a large range of possible conditions. It is a conservative approach, meaning that the elements of the solution are potentially overdesigned. However, such an approach provides a reliable solution and a required level of reliability, necessary for the functioning of water supply systems. The reliability of the bulk water supply system can be defined in terms of reliability of its storage reservoir/tank, as consumers will only notice a service interruption if the storage tank has failed (*i.e.*, run dry).

Optimization of the possible solution is based on economic and other criteria in the function of a balancing period. The balancing period of water pumping and service reservoir water balance in the urban water supply system is at least one day and may be several days, usually no more than five, (t_b = 1–5 days). A longer balance period reduces the uncertainty of solar irradiation and increases the reliability of the solution. With a longer balancing period, the system is more cost-effective from the perspective of solar energy harvesting, because the sum of overall available solar radiation is greater when the balancing period is longer. This means that the required water volume can be pumped with lower installed PV generator power. Normally, with a longer balancing period, the storage capacity of the reservoir will be higher. Therefore, regardless the fact that the daily output of water from the storage does not change significantly from day to day, due to uneven influx of water into the reservoir, it will generally be higher. The relationship between $P_{el,PV}$ and V_{op} depends on climate characteristics of the location, or on daily variability of solar radiation. With lower variability the impact on the required storage volume and the PV generator power is smaller and *vice versa*.

In the critical period approach, the estimate is mainly carried out in the following steps:

- Collecting all necessary data for estimation of water demand in the planning period, $Q_{WS(i)}$, and determining the daily water usage pattern;
- Collecting climate and other necessary data for the design of PV generator;
- Selecting the number of days for system water usage balance, *i.e.*, balancing (design) period, t_b = 1, 2, 3, 4 and 5 days;
- Selecting the most critical periods for the determination of the required power of the PV generator $P_{el,PV}$ from available time series of solar radiation $E_{S(i)}$ and water demand $V_{WS(i)}$ in accordance with the selected balancing period t_b;
- Determining the required power of PV generator $P_{el,PV}$, according to the selected critical daily balancing period t_b[10,11];

- Determining the capacity Q_{PS} and power P_{PS} of the main pumping station (MPS) for each balancing period t_b for every day i of the year [12]. The largest obtained capacity $Q_{PS(i)}$ and power $P_{PS(i)}$ is selected;

- For the selected balancing period t_b, determining the required operative reservoir volume V_{op} for the selected PV generator power $P_{el,PV}$ and the period of its work during the day (inflow), according to the foreseen regime of hourly water consumption in a settlement (outflow) for each balancing period t_b, for every day i of the year. The largest obtained volume $V_{op(i)}$ is selected;

- Determining the cell area A_{PV} (m^2), for each balancing period t_b [10];

- Analysis and ranking of obtained solutions of PV generator power $P_{el,PV}$, operative reservoir volume V_{op} and total power P_{PS} of MPS of different balancing period t_b.

The input data for the analysis are: climate data (daily average air temperature, daily average solar irradiation, average number of hours of daily insolation), water supply system configuration (water intake, pump station, service reservoir), daily and hourly water demand and fluctuation during the day. It is also necessary to determine the constraints related to the construction of a PV generator, pump station and service reservoir, as well as legal, environmental, social and other requirements.

At the beginning of the analysis it is necessary to define the daily quantity of water in a settlement V_{WS} (m^3/day), according to settlement characteristics and water consumption regime throughout the years of the planning period.

After this, the daily water usage pattern in a settlement $V_{WS,t}$ (m^3/h) in the period of $t = 1,..., 24$ hours is determined (diurnal pattern). The same pattern is used for each day i.

Based on the obtained values, the minimum required size of the PV generator is determined, which provides the necessary inflow of water in the critical period. This procedure is simple, because the relation between $P_{el,PV}$ and V_{PS} is linear.

Based on the selected/calculated initial values, $P_{el,PV}$ and V_{PS}, which satisfy water demand V_{WS} in the planning period, the minimum required $P_{el,PV}$ is determined from established differences $\Delta V_{tb,i}$:

$$\Delta V_{tb,i} = V_{PS,tb,i} - V_{WS,tb,i}\,, \; i = 1, 2, ..., 365 \tag{1}$$

The critical day/period $t^{*}_{Pel,PV,tb,i}$ for PV generator design is determined by the minimum daily difference:

$$\min \Delta V_{tb,i} \Rightarrow t^{*}_{Pel,PV,tb,i} \tag{2}$$

where $\Delta V_{tb,i}$ is an acceptable difference in practice application.

When the balancing period $t_b > 1$ day, calculation series of t_b days are performed, where series are formed with calculation step always $\Delta i = 1$ day, $i = 1, 2, \ldots, 365$. If the sum does not end with the last member of the observed series, the process ends with the next member of the same series.

Available insolation E_S, *i.e.*, electric energy $P_{el,PV}$ determines the period of the pumping station operation T_S with uniform rate during daily work period.

The required operation volume of service reservoir V_{op} is obtained by standard procedure, using a mass diagram with cumulative pumping curve plotted on it [12], by using of Excel ©2007 (Microsoft). Time step for calculation is one hour, $t = 1, \ldots, 24$ hours. The size of service reservoir has been calculated for each day i in accordance with each balancing period t_b in the year. In general, the critical day/period for the design of volume reservoir $t^{*}_{V,tb,i}$ is the day with maximum water demand, providing that on day available insolation $E_{S(i)}$ is sufficiently high. It should be noted that the fire and emergency volumes are not taken into account for this case. These volumes are more or less constant, while the operation volume is variable during the day. In accordance with the respective norms and particular situations, influence of fire and emergency volumes on operation volume analysis may be neglected [12]. However, the total required volume of service reservoir is the sum of operation, fire and emergency volumes.

The required volume V^{*}_{op} for each alternative t_b is:

$$V^{*}_{op} \geq \max V_{op,i} \tag{3}$$

2.2. The Choice of a Compromise Solution

The general objective of the system design is finding the power of the PV power plant that will, in the best manner possible, meet all consumer needs for water, with minimal construction and operation costs of the system. The requested result of the optimization process is the best compromise solution between the pairs of $P_{el,PV}$ and V_{op} and also the pump station power P_{PS} that best meets the set objectives. The optimal combination x of PV generator power $P^*_{el,PV}$ (f_1), operating service reservoir volume V^*_{op} (f_2) and power P^*_{PS} (f_3) of MPS is sought for the selected alternatives X:

$$DR[f_1(x), f_2(x), f_3(x)], x \in X \tag{4}$$

DR means to apply the appropriate decision rule(s) and to find the best-compromise solution x^* from the set of alternatives X. The standard trade-off method could, among other, be used for the selection of a compromise solution [13]. In this case, the economic criterion is dominant. However, the problem can be analyzed by extending the number of criteria, of which the reliability of water supply is the most important one.

General sustainable objectives are related to economic, social and environmental aspects of the problem. Most of the environmental objectives are fulfilled by using green energy instead of traditional. Social objectives relate to basic water service price as measurable criteria and sustainable green city environment as general commensurate criteria. The fulfillment of both sustainable objectives is closely related to economical characteristics of the solution. Nowadays, the economic criteria are still dominant and for this reason good economic analysis of the problem is the basis for the solution and alternative evaluation.

The economical approach, according to the concept of Life Cycle Cost *LCC* (€) [8,9], is developed to be the best indicator of economic profitability of the system cost analysis. *LCC* takes into account the initial capital cost $C_{capital}$, present value of replacement cost $C_{replacement}$ and present value of maintenance cost $C_{maintenance}$:

$$LCC = C_{capital} + C_{replacement} + C_{maintenance} \tag{5}$$

In the case of urban water supply systems, the economic objective is to minimize possible economic losses that occur due to not using conventional energy sources which are still cheaper than green sources. These losses are expected to decrease over the time, because PV generators are becoming cheaper and conventional energy more expensive [14,15].

In this paper we will simplify the economic analysis, providing sufficient information for understanding the problem characteristics.

3. EXAMPLE

This paper presents a hypothetical example of a settlement with a population equivalent of 8970. The settlement is located on an island in the southern Mediterranean part of Croatia. It is in a hilly area of the island and has one water reservoir located at a ground elevation of 259 m above sea level. Water flows into the reservoir from the wet basin of the pump station. Water flows into the wet basin of the pump station by gravity from the spring. Total head of the pump station is 82.41 m. The water quality is satisfactory and does not need treatment. The positions of the basic facilities of the water supply system are shown in Figure 3 [16].

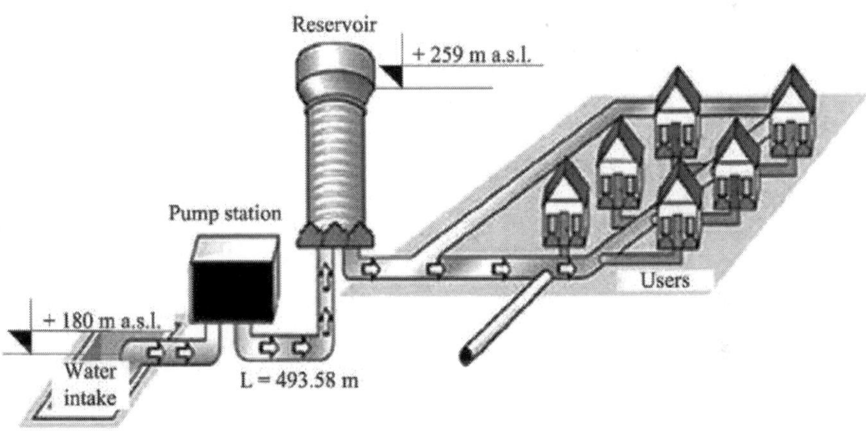

Figure 3. Case study schematic layout.

The analysis is conducted according to the presented methodology. Specific water consumption per capita q_{sp} is 160 L per day. The daily water consumption pattern through the year is shown in Figure 4 [12].

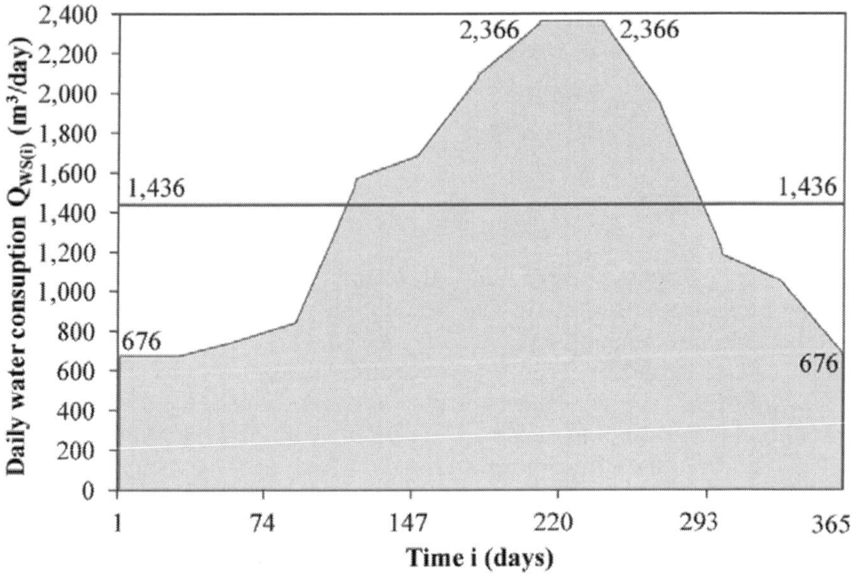

Figure 4. Daily water demand during the year.

Hourly water consumption pattern [12] in the settlement is determined by the daily regime of consumption, as shown in Figure 5.

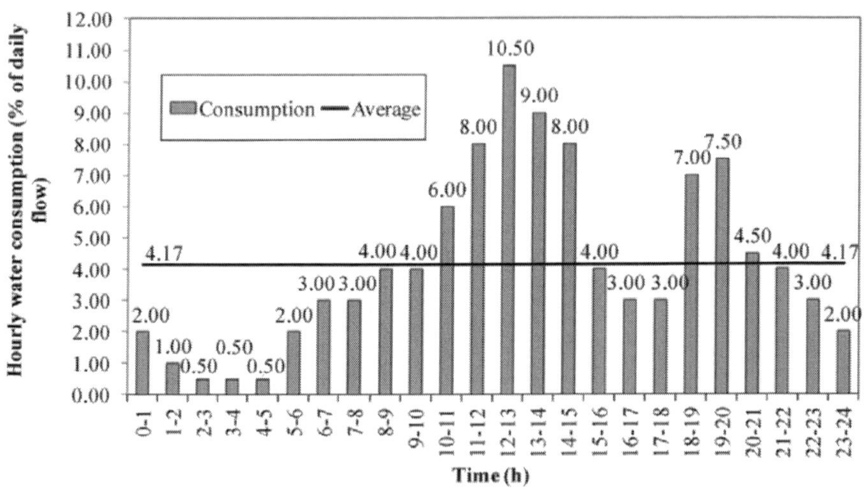

Figure 5. Hourly water demand during typical day.

For the considered location, the average pump head is H_{PS} = 82.41 m, average efficiency of the inverter and motor pump unit is η_{MPI} = 0.75, cell temperature coefficient is α_C = 0.005 °C^{-1} and temperature of the PV generator in Standard Test Condition is T_0 = 25 °C.

The calculation is made with 50% of an average array output in relation to rated output (η_S = 0.5). This approximation is taken because the calculation from data in [3] shows that for the southern Mediterranean part of Croatia, the ratio of surface size of the daily profile of E_S during T_S and surface size of the daily profile of E_S during daily duration of solar radiation is about 50% in the winter period of the year. In the summer period this ratio is even up to 75%. However, the relevant value is 50%, because the critical day/period $t^*_{Pel,PV,tb,i}$ for the PV generator is in winter.

The average daily global radiation $E_{S(i)}$ and average daily insolation period $T_{S(i)}$ are shown in Figure 6, while the average daily cell temperature $T_{cell(i)}$ and average daily ambient air temperature $T_{a(i)}$ are shown in Figure 7 [17].

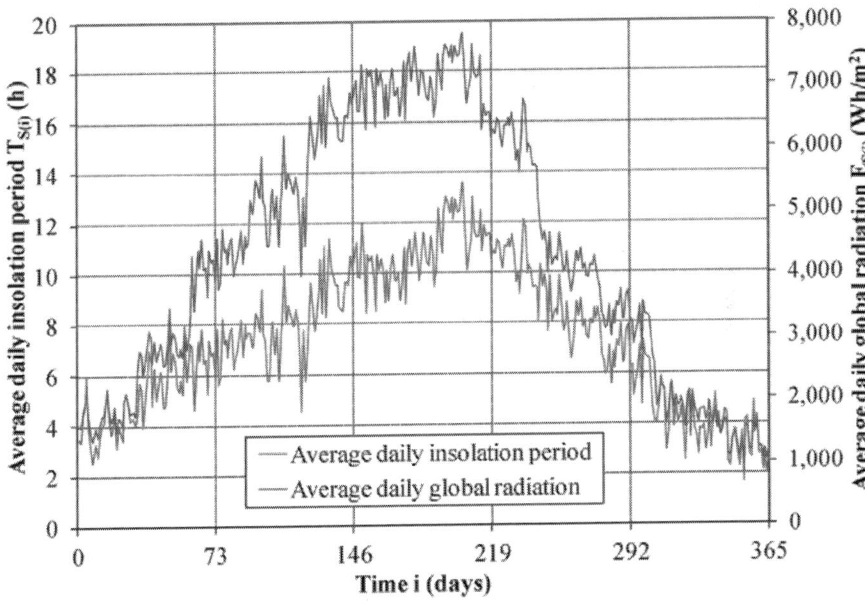

Figure 6. Average daily solar insolation and duration of sunlight.

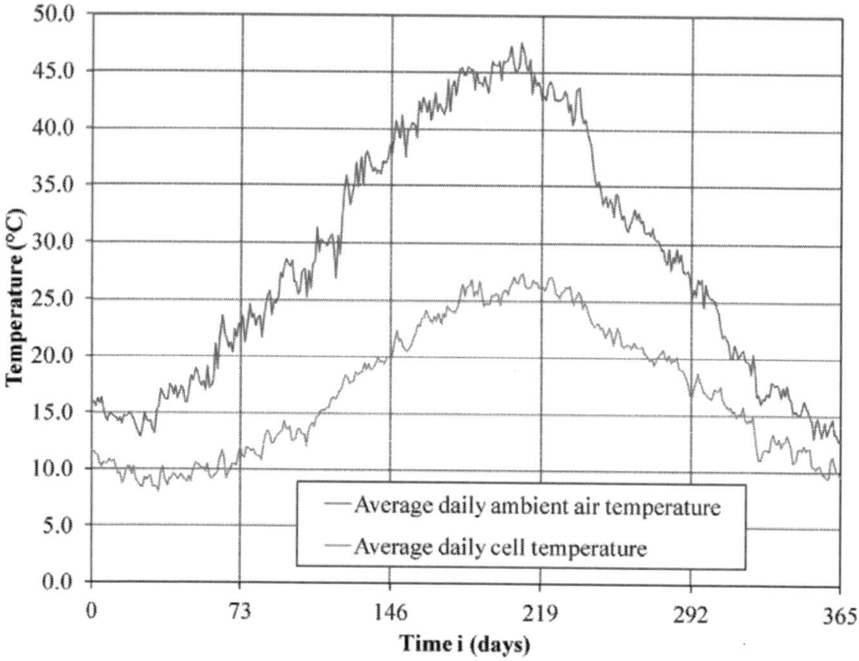

Figure 7. Average daily cell temperature and average daily ambient air temperature.

4. RESULTS AND DISCUSSION

Based on the given data, by applying Equation (2), critical days $t^*_{Pel,PV,tb,i}$ and $t^*_{V,tb,i}$ in the given data sets have been determined (Table 1). Also, by applying the previously presented methodology, the required PV generator power $P^*_{el,PV}$, required area of the PV generator A_{PV} (using the value of η_{PV} = 0.15) and required reservoir volume V^*_{op} (by applying Equation (3)) are calculated (Table 1 and Figure 8).

The required capacities of pump station Q^*_{PS} and power of pump station P^*_{PS} are calculated by using the value of η_{PS} = 0.90 [12] (Table 2).

By using data from literature [15,18,19,20,21,22] and Equation (5), Life Cycle Cost analysis *LCC* has been made and presented in Table 3 and Table 4.

Table 1. The length of balance period in accordance with critical days and required area for a PV generator.

Balancing periods t_b (days)	1	2	3	4	5
Critical days $t^*_{Pel,PV,tb,i}$ (days in year) for the $P^*_{el,PV}$	352	344–345	344–346	349–352	348–352
Critical days $t^*_{V,tb,i}$ (days in year) for the V^*_{op}	244	244–245	243–245	243–246	242–246
Required area for PV generator A_{PV} (m^2)	3417	3185	2959	2799	2690

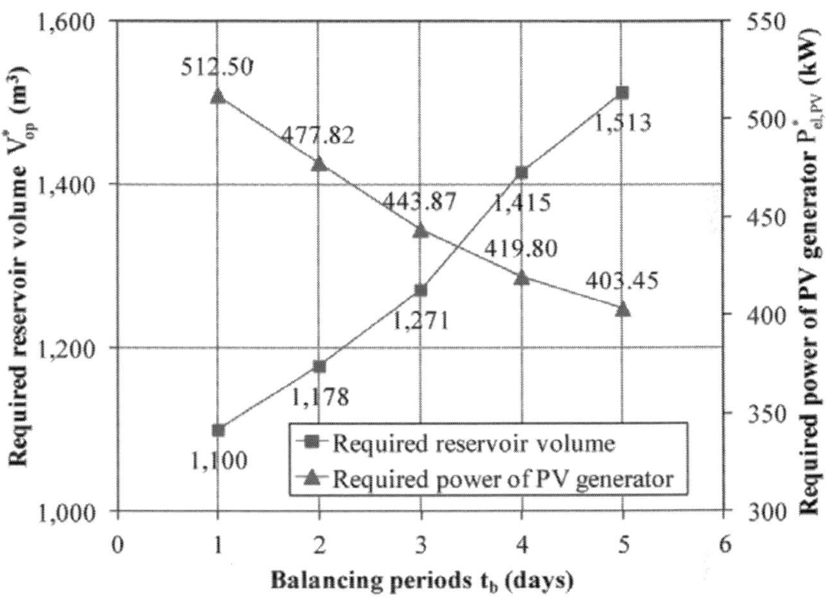

Figure 8. The length of the balance period in accordance with the required power of a PV generator and required reservoir volume.

Table 2. Capacity Q^*_{PS} and power P^*_{PS} of pump station for different balancing periods t_b.

Balancing periods t_b (days)	1	2	3	4	5
Capacity of pump station Q^*_{PS} (L/s)	118	133	124	117	111
Power of pump station P^*_{PS} (kW)	106.00	119.47	111.39	105.10	99.71

Table 3. Costs and lifetime aspects of the system components.

Component	Unit cost, c (mean value of literature data)	Maintenance costs in the first year, k (%)	Lifetime, L_P (years)	Real interest rate, k_d (%)	Inflation rate (%) f_0	Inflation rate (%) f_1
PV generator	1.5 (€/W)	1	25	8	4	4
Invertor	0.5 (€/W)	0	10	8	4	4
Service reservoir	400 (€/m^3)	1	25	8	4	4
Pump station	1 (€/W)	3	15	8	4	4

Table 4. Costs of obtained solution variants.

Bal. periods t_b (days)	Power $P^*_{el,PV}$ (kW) of PV	Volume V^*_{op} (m^3) of SR	Power P^*_{PS} (kW) of PS	$C_{capital}$, Total (€)	C_{repl}, Total (€)	$C_{(O\&M)}$, Total (€)	LCC (€)
1	512.50	1,100	106.00	1,570,996	554,084	263,418	2,388,498
2	477.82	1,178	119.47	1,546,310	526,705	269,198	2,342,213
3	443.87	1,271	111.39	1,507,525	489,479	261,567	2,258,572
4	419.80	1,415	105.10	1,510,697	462,815	260,740	2,234,252
5	403.45	1,513	99.71	1,511,808	444,154	259,436	2,215,398

It is evident that the fifth alternative is the most cost-effective. As a matter of fact, the difference in the maximum and minimum value of *LCC* is less than 8%, so a suitable compromise solution can be selected among them. We should not forget that the downward trend in prices of a PV generator (price reduction by the day by 4% to 7% annually within the observed period of N = 25 years) is very significant [15]. This means that construction costs of the reservoir will have a decisive role.

The service reservoir, which has the function of balancing the inflow and supply of water to towns also acts as energy storage and thus determines the size of the required power of the PV generator. Theoretically assumed dependence of the size of conjugate pairs of $P_{el,PV}$ and V^*_{op} as the basis for solution optimization was confirmed by the obtained results. It can thus be seen that, by using a longer balancing period t_b, the power of the PV generator is becoming smaller and the volume of service reservoir is becoming bigger. This functional dependence, in function of the length of the balancing period and electric/water energy storage, is the basis for the optimization of the system. Longer balance periods would probably yield more favorable solutions, considering that the PV generator is the most expensive element, although not significantly.

The system produces a surplus of electric energy during the sunshine period in the year, since it is designed to satisfy the needs in the critical period (Figure 9). With greater power of the PV generator, the available surplus is also higher. If there is a connection with the local grid (system), the aforementioned surplus of electric energy could be distributed to other consumers or can be sold.

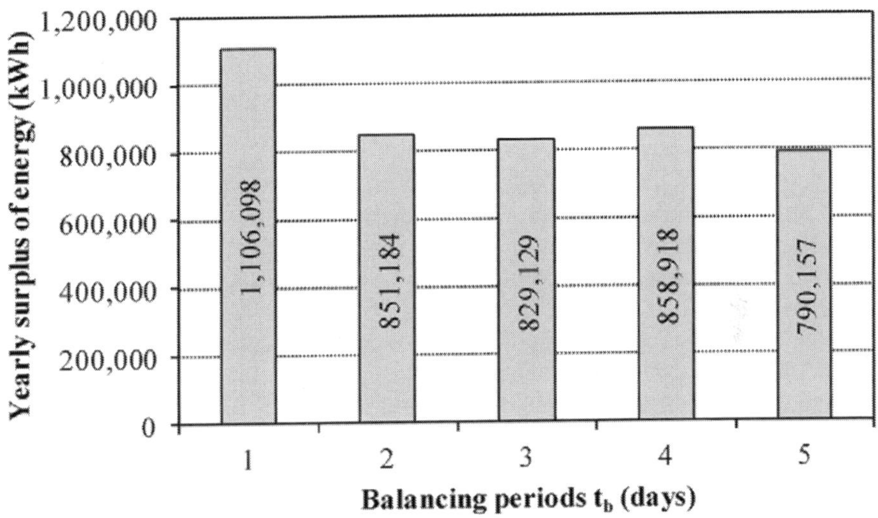

Figure 9. Yearly surplus of energy from a PV generator in relation to the pump station needs.

The obtained profit from the sale of the produced electric energy surplus profit (€) and net cost (€) over the observed period of N = 25 years are obtained by using the net present value (*NPV*) [10] (Table 5). In this calculation, the sale price of electric energy from the PV generator in Croatia is 0.07 €/kWh [23] and the nominal discount rate for PV electric energy is 10% [24]. Net cost is the difference between *LCC* and profit. It can be seen that *LCC* is decreasing from 28% for t_b = 5 days, up to 37% for t_b = 1 day. This can significantly contribute to the economic benefits of the proposed solution if there is a market for green energy. Potential revenue from the sale of green electrical energy surplus changes the selection of obtained solution alternatives, considering only the economic criterion. In this case, alternative 1 is the most economical.

Table 5. Profit and net cost of the system over the observed period of N = 25 years.

Balancing periods t_b (days)	1	2	3	4	5
Profit (€)	878,512	676,049	658,530	682,185	627,578
Net cost (€)	1,509,986	1,666,164	1,600,042	1,552,067	1,587,820

In the presented climate area daily surpluses are higher in the summer than in the winter period. Such behavior of the system is particularly beneficial in tourist areas of the Mediterranean which are characterized by a significant increase in water consumption during the summer because of the tourists.

The biggest problem for operation of the proposed system is uncertainty of daily solar irradiation due to possible cloudiness. However, by using the critical period approach, significant system capacity reserves are achieved and, consequently, the reliability of the system work is increased. The selected power of the PV generator is able to cover uncertainty in water demand in the summer period, since the critical period for design is the winter low irradiation period (maximum PV generator power), while the selected volume of service reservoir is able to cover uncertainty in winter solar irradiation, since the critical period for volume design is the summer peak water demand (maximum reservoir volume). By using longer balancing period t_b the system reliability will be higher.

The advantage for having the system in combination with a network of other energy sources is the opportunity to provide redundancy in the system. Also, the advantage of the use of the solar energy system is that it provides energy into the grid and itself provides some measure of overall redundancy.

This concept of energy supply operation does not generate any atmospheric emission of greenhouse gases, which contributes significantly to reduction of the impacts of climate changes, as well as to meeting the requirements arising from relevant strategies [25]. A photovoltaic installation makes no noise, making it suitable for construction close to people.

Long-term goals of the water sector include continuous adaptation to climate change and reducing the industry carbon footprint. With greater power of a PV generator, the reduction of greenhouse gas emissions is also greater. If it is assumed that all electric energy is produced in a coal-fired power plant, which typically emits 0.95 kg of CO_2/kWh [11], a PV generator would then reduce CO_2 emission over the observed period of N = 25 years (Table 6).

Table 6. Amount of reduced CO_2 emission over the observed period of N = 25 years.

Balancing periods t_b (days)	1	2	3	4	5
Amount of reduced CO_2 emission (t)	33,120	30,879	28,685	27,129	26,072

This type of solution for electric energy supply of water supply systems also has other environmental benefits, as it reduces the need for central power station capacity and the need to reinforce transmission and distribution systems. It also provides electricity at the point of use so that energy losses should be much lower than when power is transmitted over a distance of several kilometers.

The weakness of the obtained solution can be seen due to the large surface area required for installing solar panels, which is not always easy to find in urban areas. In line with the expected increase of the PV cell efficiency (from the present 15% up to the expected 30% in the next 30 years [26], where in laboratory conditions 44% of the PV cell efficiency has already been achieved [27]), the required area for installing of solar panels will be reduced proportionally to the increasing of their efficiency. Therefore, this concept will be more suitable for the use in urban areas.

The adoption of this concept may contribute to the EU 20-20-20 target by 2020 [25] and also may increase revenue through the Renewable Obligation Certificates, ROCs. It is obvious that the proposed solution is promising, but its use will greatly depend on the characteristics of local climate, PV systems, water supply systems and also of the community commitment to water and energy supply sustainability. The biggest threats in using the presented concept could be mistrust and skepticism from managers and engineers of water supply companies. Despite the fact that the presented concept is technologically very simple and straightforward for application, it is different from the usual concept of water supply. It is therefore necessary to continue with researching and developing the proposed solutions, as well as with publishing the new obtained results.

5. CONCLUSIONS

It has been shown that the proposed concept of the use of a PV generator for electric energy supply of the main pump station in a bulk water supply system is feasible. By using PV electric energy for water pumping in urban water supply system, as has been presented, this system becomes a "green water system." Given that the primary energy source (the Sun) is inexhaustible and free of charge, energy supply, essential for the operation of water supply system, is sustainable.

The obtained results show that the proposed solution can reliably and continuously supply the main pump station with green electricity, providing a reliable and continuous water supply to meet the needs of consumers. In accordance with the proposed, the reliability of such a system is high.

The proposed solution can be used as a stand-alone power system or in combination with electric energy from the regional network (fossil fuels fired power plant). It can be concluded that the solution analyzed in this study would be very useful in all situations where the supply of electricity from energy power system is limited or when it is necessary to increase the capacity of a long transmission system (*i.e.*, isolated locations like islands). The proposed solution is also convenient to apply for peak power generation.

It is obvious that there is a whole range of possible applications, and the solution itself will depend on the characteristics of the location and the problem addressed. In any case, the proposed concept of energy production and use provides opportunities to support the sustainability of cities or a smart city environment [28].

ACKNOWLEDGEMENTS

We would like to thank Ms Marina Culic for reviewing of English language and style in this manuscript.

AUTHOR CONTRIBUTIONS

First author has expanded and elaborated the concept and the idea of the second author about the manuscript content. Both authors were involved in preparing of the manuscript.

REFERENCES

1. The Priority Actions Programme Regional Activity Centre (PAP/RAC): Integrated Coastal Urban Water System Planning in Coastal Areas of the Mediterranean—Volume I, Principles and Planning. PAP/RAC: Split, Croatia, 2007.

2. Matić, Z. *Solar Radiation on Area of the Republic Croatia*; Energetic Institute Hrvoje Požar: Zagreb, Croatia, 2007. (In Croatian)
3. SODA—Solar Radiation Data—Solar Energy Services for Professionals. Available online: http://www.soda-is.com/eng/services/services_radiation_free_eng.php (accessed on 9 January 2013).
4. Varkom PLC. *Water Consumption Data*; Varkom PLC: Varaždin, Croatia, 2013. (In Croatian)
5. Chen, H.; Cong, T.N.; Yang, W.; Tan, C.; Li, Y.; Ding, Y. Progress in electrical energy storage system: A critical review.*Progr. Nat. Sci.* **2009**, *19*, 291–312.
6. Electricity Storage Association (ESA)—Technology Comparisons. Available online: http://www.electricitystorage.org/technology/tech_archive/technology_comparisons (accessed on 4 December 2013).
7. Moreira, D.F.; Ramos, H.M. Energy Cost Optimization in a Water Supply System Case Study. *J. Energy* **2013**, *2013*, 1–9.
8. Bakelli, Y.; Hadj, A.A.; Azoui, B. Optimal sizing of photovoltaic pumping system with water tank storage using LPSP concept. *Sol. Energy* **2011**, *85*, 288–294.
9. Ghoneim, A.A. Design optimization of photovoltaic powered water pumping system. *Energy Convers. Manag.* **2006**, *47*, 1449–1463.
10. Kenna, J.; Gillett, B. *Solar Water Pumping: A Handbook*; Intermediate Technology Publications: London, UK, 1985.
11. Margeta, J.; Glasnović, Z. Exploitation of temporary water flow by hybrid PV-hydroelectric plant. *Renew. Energy* **2011**,*36*, 2268–2277.
12. Margeta, J. *Water Supply: Planning, Design, Management and Water Purification*; Faculty of Civil Engineering and Architecture, University of Split: Split, Croatia, 2010. (In Croatian)
13. Chankong, V.; Haimes, Y. *Multiobjective Decision Making: Theory and Methodology*; Elsevier-North Holland: Washington, DC, USA, 1983.
14. Campoccia, A.; Dusonchet, L.; Telaretti, E.; Zizzo, G. Comparative analysis of different supporting measures for the production of electrical energy by solar PV and Wind systems: Four representative European cases. *Sol. Energy* **2009**,*83*, 287–297.

15. SolarServer—PVX spot market price index solar PV modules. Available online: http://www.solarserver.com/service/pvx-spot-market-price-index-solar-pv-modules.html (accessed on 8 January 2014).
16. District of Columbia Water and Sewer Authority—About Drinking Water Quality in Washington, DC, USA. Available online: http://www.dcwater.com/drinking_water/about.cfm (accessed on 10 December 2013).
17. Meteorological and Hydrological Service (MHSC). *Climatic data for Palagruža and Komiža — From 1995. till 2006*; MHSC: Zagreb, Croatia, 2007. (In Croatian)
18. Four Peaks Technologies, Inc.—Solar electricity costs. Available online: http://solarcellcentral.com/cost_page.html(accessed on 2 January 2014).
19. Hidroing Ltd. *The Regional Water Supply System: Neretva, Pelješac, Korčula, Lastovo, Mljet—Novelation of Pre-Investment Programe*; Hidroing Ltd.: Split, Croatia, 2004. (In Croatian)
20. Hrvatske Vode PLC. *Water Supply Plan of Split and Dalmatia County*; Hrvatske Vode PLC: Split, Croatia, 2008. (In Croatian)
21. Penstar Ltd. *Price Book*; Penstar Ltd.: Cardigan, UK, 2012.
22. White International. *CNP Industrial Pumps—Price List 2011*; White International: Mippera, Australia, 2012.
23. Croatian Energy Regulatory Agency (HERA). *Methodology for Determining the Amount of Tariff Items for Electricity within the Universal Service*; HERA: Zagreb, Croatia, 2013. (In Croatian)
24. U.S. Department of Energy. *The Prospect for $1/Watt Electricity from Solar. $1/W Workshop*; U.S. Department of Energy: Washington, DC, USA, 2010.
25. European Commission. *Energy Roadmap 2050*, 2012. Available online: http://ec.europa.eu/energy/publications/doc/2012_energy_roadmap_2050_en.pdf (accessed on 19 December 2013).
26. Goetzberger, A.; Luther, J.; Willeke, G. Solar cells: Past, present, future. *Sol. Energy Mater. Sol. Cells* **2002**, *74*, 1–11.
27. The National Center for Photovoltaics (NCPV). Research Cell Efficiency Records. Available online: http://www.nrel.gov/ncpv/images/efficiency_chart.jpg (accessed on 16 April 2014).

28. United Nations. Report of the United Nations Conference on Sustainable Development. In Proceedings of the United Nations Conference on Sustainable Development, Rio de Janeiro, Brazil, 20–22 June 2012.

2
CHAPTER

SOLAR DRYER APPLICATION FOR OLIVE OIL MILL WASTES

Irene Montero *, María Teresa Miranda, Francisco José Sepúlveda, José Ignacio Arranz, Carmen Victoria Rojas and Sergio Nogales

Department of Mechanical Engineering, Energy and Materials, Industrial Engineering School, University of Extremadura, Av. Elvas s/n, 06006 Badajoz, Spain

ABSTRACT

Global waste production has raised recently due to numerous agricultural and industrial activities. Among other actions devoted to waste reduction, revaluation seems to be the most advantageous one from an environmental and economic point of view. In the olive oil sector, by-product management (namely olive pomace, olive mill wastewater, and sludge residue) poses serious problems for companies, with the energy use of these wastes being a feasible option to solve these problems. Due to their high moisture content, drying constitutes the main stage for a possible bio-fuel conversion. This research work deals with the analysis of drying for the three main wastes from olive oil by using a prototype dryer. This equipment has different working regimes depending on the kind of convection (natural or forced for passive and active mode, respectively), the incidence of solar radiation on the product (indirect or mixed type) and the use of an additional system for energy supply (hybrid type). From the results, it could be said that all the wastes were suitable for solar drying. However, drying conditions were different for each case. Olive pomace and olive mill wastewater showed promising results for solar drying application in mixed active mode. Concerning sludge residue, its special physical structure could employ the use of turners for drying, in order to improve heat transfer to the product. Moreover, hybrid active mode enabled a considerable reduction in drying time, being an aspect to take into account for its use during low solar radiation or at night time.

Keywords: drying kinetics; thin layer; olive pomace; olive mill wastewater; sludge residue

1. INTRODUCTION

In a global context where waste production increases constantly, its revaluation is becoming more and more important, from an environmental and economic point of view, implementing new measures and procedures that ensure the correct use of by-product recycling is necessary. In order to attain this goal, the use of wastes as fuels or other ways for energy generation is an essential part of sustainable strategies.

The olive oil industry is very important in the worldwide economy. Mediterranean countries produce over 98% of olive oil in the world (around 2.5 million tons per year), increasing 40% in the last decade [1].

Spain is the largest olive oil producer and exporter in the world (48%) [2], way ahead of Italy, which ranks second (21%). Some regions, such as Andalusia, Castilla-La Mancha and Extremadura, located in the south of Spain, represent 93% of national production, with over 2 million hectares of olive grove.

For olive processing, two techniques are currently used: the traditional system or three-phase extraction system, with which by-products such olive pomace (OP) and olive mill wastewater (OMW) are obtained; and the two-phase extraction system, whose main waste is sludge residue (SLG). The difference in material balance for these procedures might be summarized in water addition reduction in two-phase extraction system, with the resulting increase in solid waste generation (SLG) and the considerable reduction in liquid waste (OMW). Material balances of the oil mill by-products resulting are shown in Table 1 [3].

Table 1. The input-output mass balance (kg) of the olive oil process.

Olive processing	Input products		Output		
			Pomace	Wastewater	Oil
A three-phase process	Olive paste	100	-	-	-
	Process water	90	-	-	-
	Total	190	62	115	13
A two-phase process	Olive paste	100	-	-	-
	Process water	15	-	-	-
	Total	115	87	15	13

Olive pomace is a solid waste composed of pulp, stone, waste oil and water in varying proportions depending on the technology. Usually, olive pomace is processed again by extractors in order to obtain waste oil, generating a new waste with less moisture and higher heating value that is suitable for biofuel use [4]. Olive mill wastewater is a liquid waste that comes from water, contained in olives itself and washing process. It is blackish and usually contains small quantities of oil and pulp. It constitutes a very pollutant waste, due to its high BOD (biochemical oxygen demand) and COD (chemical oxygen demand) levels [5]. Sludge residue is a semi-solid compound made of olive mill wastewater and olive pomace, produced in high quantities and causing environmental impact. It has high moisture percentage, which makes suitable transport and storage systems necessary.

Even though two-phase technology aimed to reduce environmental problems concerning traditional systems, its massive implementation has caused a considerable increase in sludge residue production.

In this regard, the management of the above-mentioned wastes implies a serious problem for olive oil industries, because of their high environmental pollution and the expensiveness of their treatment [6]. Waste revaluation, along with its positive effect on the environment, would enable significant operating and economic improvements for olive oil companies.

Among the different treatments and removal techniques of these wastes (anerobic digestion, composting, centrifugation, floculation, *etc.*), their use as a soil improver, with or without composting, is included among the most used options in practice, because of their beneficial effects, simplicity, and low costs [7,8]. Thermal procedures also represent one of the most used options in practice, with drying as the first treatment in order to revaluate wastes as biofuels.

Drying kinetics of wastes in olive oil industry has been studied by several researchers, by using fluidized beds [9], convective dryers [10], microwave-convective dryers [11] and solar dryers of different kinds (passive, active and hybrid) [12,13,14].

Solar drying is considered a technique with high potential for by-product reuse. Although solar drying has been used for food products drying, its use for wet by-products has several advantages.

Passive mode is the simplest kind of operation, with the lowest initial investment. However, active mode enables more independence from environmental conditions, allowing a more effective drying process. Finally, hybrid mode makes higher variable control possible, as well as it works with low isolation [15].

In the south of Spain, the use of renewable energy sources, such as solar energy, plays an important role in the drying of these kind of wastes, due to the concurrence of two important factors: the abundance of olive oil industries (availability of by-products) in areas with the highest solar radiation in Europe.

Currently, one of the most important options for energy use of biomass wastes consists of pelletizing and its subsequent application in industrial boilers. There are some research works that analyze densification of wastes from the olive oil industry, with satisfactory results, although their high moisture could imply high costs and therefore a problem to be solved [16,17]. Drying kinetics studies might contribute to reducing this cost and improving the efficiency of pelletizing.

For all those reasons, the use of solar drying for by-products from the olive oil industry might imply a viable option in order to revaluate wastes for their use as bio-fuels.

In the present research work, different kinds of solar drying are analyzed for the three above-mentioned wastes, with the aim of assessing their possible technical implementation.

2. MATERIALS AND METHODS

The analyzed wastes came from two oil mills located in Extremadura region, in the south-west of Spain. OP and OMW samples were obtained from a three-phase extraction system, whereas SLG samples were obtained from a two-phase extraction system. Table 2 shows the physical-energy characterization for the wastes.

Table 2. Physical-energy characterization. wb, wet basis; db, dry basis.

Property	Wastes		
	Olive Pomace (OP)	Olive Mill Wastewater (OMW)	Sludge Residue (SLG)
M (% wb)	50–58	88–92	68–72
BD (kg/m^3 wb)	703	1022	1065
VM (% db)	83.50	78.27	80.18
FC (% db)	12.17	11.41	15.58
Ash (% db)	4.33	10.32	4.24
HHV (MJ/kg db)	21.18	25.14	28.14
LHV (MJ/kg db)	17.26	7.50	22.26

Whereas moisture is the main parameter that was analyzed in this paper, it was interesting to include some of the most significant physical-energy properties of the wastes, especially when one of the objectives is the energy use of by-products.

The dryer, described in full detail by Montero *et al.* [18] and installed in University of Extremadura (Badajoz, Spain), has a flat-plate solar collector, a drying chamber for two trays with product, an energy supply system and a chimney for exhaust air, as well as different auxiliary elements for operating, control, and measuring. The dryer works as follows: the collector picks up solar radiation, heating the air and reducing its relative humidity. Afterwards, the air enters the drying chamber, where it is put in contact with the products and reduces their moisture. Finally, the air goes out through the chimney with less temperature and higher relative humidity.

Figure 1 shows the equipment: 1—temperature and relative humidity thermocouple; 2—speed sensor; 3—product temperature thermocouple; 4—resistors; 5—balance; 6—fan; 7—folding hood; 8—movement wheels; 9—drying tray; 10—lock system; 11—insulation; 12—absorbing surface; 13—polycarbonate; 14—folding glass doors; 15—control panel; 16—chimney; 17—support for radiation simulator; 18—weather station; 19—transmission and control system; 20—personal computer. The equipment has six working regimes (Table 3), depending on the kind of convection (natural or forced for passive or active mode, respectively), the absence or incidence of solar radiation on the product (indirect or mixed types, respectively) and the use of additional systems to supply energy (hybrid type).

In passive mode, the air moves naturally as a consequence of density variation, due, in turn, to temperature changes. However, in active mode, forced convection is provoked by the use of a fan located in the upper part of the drying chamber.

On the other hand, indirect, mixed and hybrid types are based on different combinations of the following options: air heating in the collector, direct solar radiation on the product and the use of resistors in order to improve air heating.

In active mode, the speed of the fan was fixed to generate two different air mass flows, $q1$ and $q2$, with 0.22 and 0.24 kg/s, respectively. In previous studies, devoted to startup dryer optimization [12], the first value was considered to be optimum for active mode operation in this equipment, reaching the highest air temperature within the drying chamber.

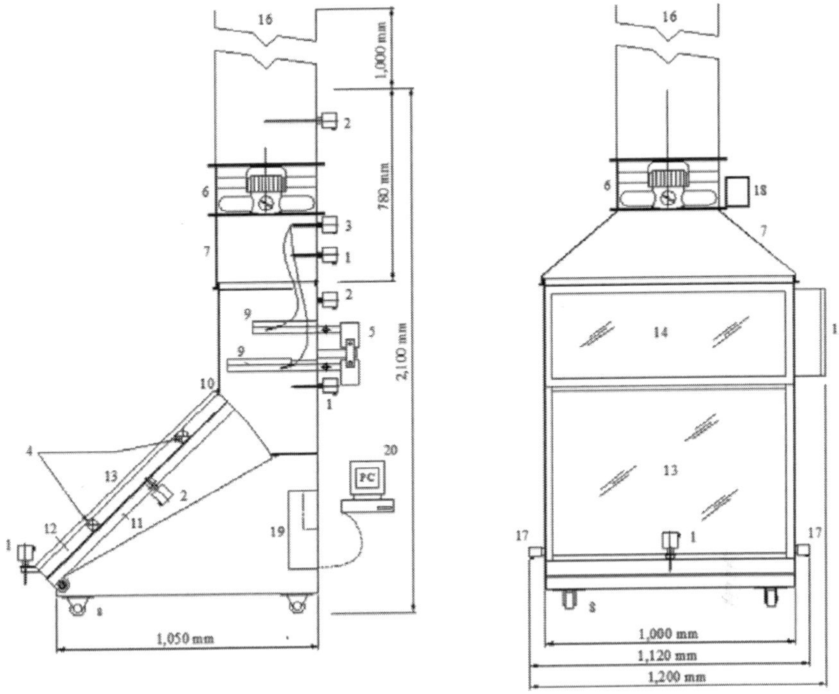

Figure 1. Solar dryer.

Table 3. Operating modes of the solar dryer. √, on; -, off.

Mode	Type	Pre-Heating	Solar Radiation	Resistors
Passive	Indirect	√	-	-
	Mixed	√	√	-
	Hybrid	√	√	√
Active	Indirect	√	-	-
	Mixed	√	√	-
	Hybrid	√	√	√

The assays with hybrid type were carried out with the two 1-kW resistors working continuously.

Different assays were done for each working type of the equipment. In spite of climatological factors, the assays were carried out under similar ambient conditions, as much as possible (from May to September), monitoring the process by using a meteorological station (WatchDog Data Logger 450, Spectrum Technologies, Inc., Aurora, IL, USA). The assays were done in triplicate for each waste and operating type.

For all cases, 2 kilograms of each product was uniformly placed (forming 20–40 mm layers) on stainless steel trays (perforated for OP and non-perforated for OMW and SLG), monitoring certain variables such as weight loss, dry air temperature, relative humidity in air, product temperature, global radiation on the horizontal surface of the collector, and wind speed, among others.

The assays lasted until moisture in products was 20 ± 0.50 (% wb), for all cases.

Moisture determination was carried out through the following equations:

$$M\ (\mathrm{wb}) = \frac{m_w}{m_s + m_w} \cdot 100\ (\%\ \mathrm{wb}) \tag{1}$$

$$M\ (\mathrm{db}) = \frac{m_w}{m_s} \cdot 100\ (\%\ \mathrm{db}) \tag{2}$$

Given the different moisture levels in the three wastes, and with the aim of comparing drying kinetics regardless of initial moisture, in the present work moisture ratio (*MR*), defined as the quotient between moisture content at any time (M_t) and initial moisture (M_0), was used, according to Equation (3).

$$MR = \frac{M_t}{M_0} \tag{3}$$

Finally, efficiency levels depending on the operating mode (passive, mixed active, and hybrid active) of the dryer, were measured by Equations (4)–(6), respectively [19,20].

$$\eta = \frac{(W_0 - W_t) \cdot L}{I_t \cdot A_c} \tag{4}$$

$$\eta = \frac{(W_0 - W_t) \cdot L}{I_t \cdot A_c + E_f} \tag{5}$$

$$\eta = \frac{(W_0 - W_t) \cdot L}{\left(I_t \cdot A_c + E_f\right) + E_a} \tag{6}$$

Data corresponding to drying kinetics of OP were analyzed by Montero *et al.* [18], being the basis for the present research work.

3. RESULTS AND DISCUSSION

3.1. Initial Tests

Drying is a fundamental stage for waste revaluation as a biofuel. High moisture content implies that some heat from combustion is used to evaporate water in biofuels, reducing the useful energy. Lower moisture allows a higher flame temperature (with better temperature gradients and heat transfer), as well as shorter residence time in the combustion chamber [21].

Therefore, combustion yield in biomass boilers increases when moisture decreases. The drying process contributes to reducing transport costs [22].

Energy requirements for waste and by-product drying depends on their nature, and it can be assessed by measuring initial and final moisture. In addition, each product has a certain drying rate and maximum temperature ranges (if it is used in food) [15].

Drying kinetics of by-products is dependent on several parameters, both external (ambient temperature and humidity, solar radiation, *etc.*) and internal (temperature and humidity inside the drying chamber, flow mass, and speed of air, among others).

An increase in dry air temperature, with the subsequent decrease in relative humidity, causes a considerable reduction in drying time [23]. In turn, concerning solar dryers, an increase in solar radiation causes higher temperatures in drying chambers and, therefore, shorter operating times. Finally, the increase in air speed through the product reduces drying ratio, with a limit value for this speed from which drying times do not decrease significantly [24].

The first assays were carried out in active mode (forced convection, $q1$), with the aim of comparing the suitability of hybrid and indirect types. The obtained results are shown in Figure 2 and Figure 3.

In this way, a significant improvement in drying times can be observed when mixed type was used for the three by-products (43%, 64% and 25% for OP, OMW and SLG, respectively). Therefore, this option is preferable for those wastes set aside for energy use, whose quality parameters are not important (for instance, products for food) [25,26,27].

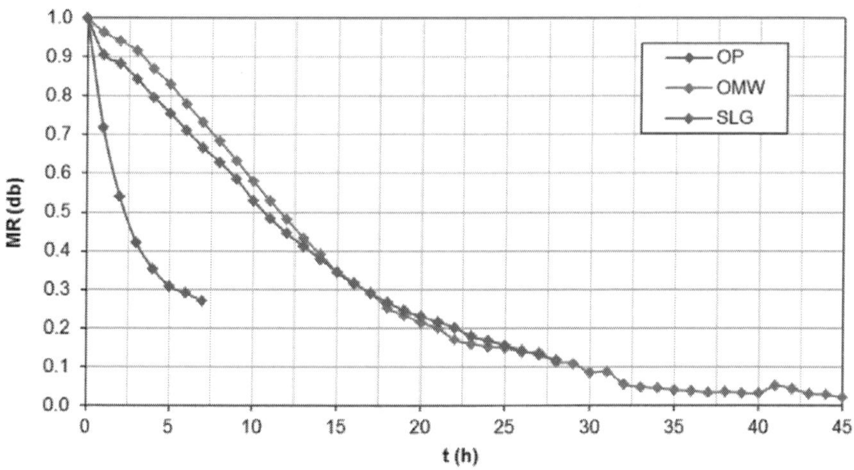

Figure 2. Moisture ratio (*MR*) (db) indirect active mode.

Even though OMW and SLG wastes showed similar results for indirect active mode, there were some differences for mixed type. After 5 h of the assay and a moisture ratio value of 0.6 (db), a reduction in drying time for OMW waste can be observed, despite its initial high moisture. This circumstance could be due to heterogeneity and the particular physical structure of SLG (that is a mixture of several products) that promotes sugar caramelization and the subsequent ball formation that occludes water, avoiding total water evaporation [28].

Once the advantages of mixed type *versus* indirect types were observed, a comparison between passive and active operating modes was carried out. Thus, Figure 4 shows moisture ratio evolution over time for mixed passive mode.

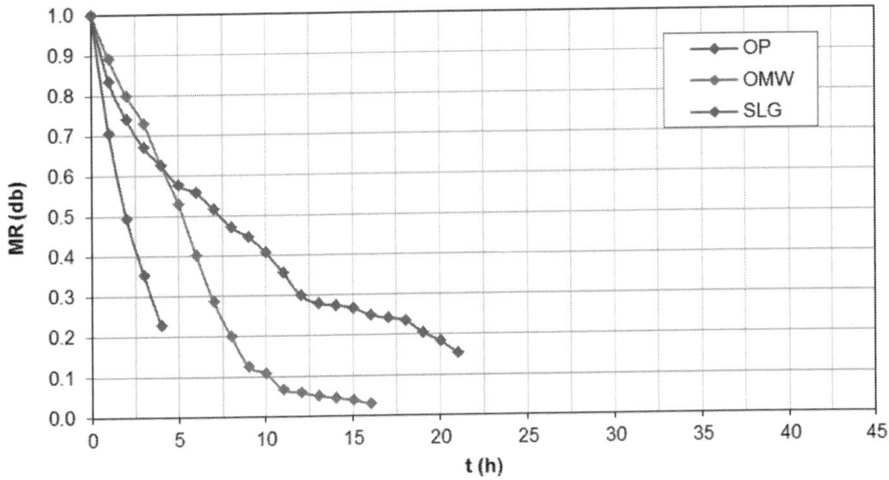

Figure 3. *MR* (db) mixed active mode.

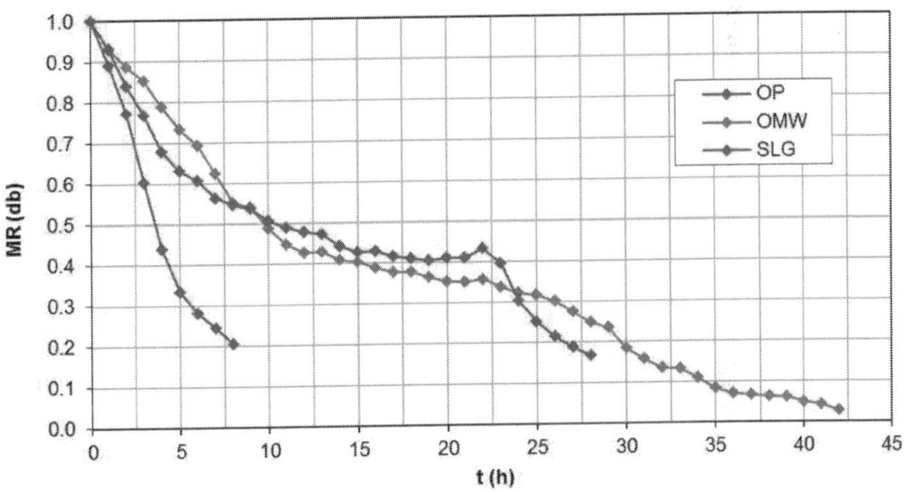

Figure 4. *MR* (db) mixed passive mode.

Shorter drying times were obtained for OP, although these values were higher than the obtained ones for mixed active mode (Figure 3). SLG showed shorter drying times than OMW (even though it experienced reabsorption at the 22nd hour of the assay), being the opposite for mixed active mode. This could be due to the lack of control over operating conditions in the chamber for natural convection, which made the drying process worse for OMW (on account of its particular physical structure) and better for SLG.

In this way, through active mode, a considerable reduction in drying time (compared to passive mode) was obtained, improving values for OMW wastes as this situation increased moisture removal.

In all cases, drying time was shortened, in mixed active mode, between 25% and 62%.

3.2. Influence of Air Mass Flow

Figure 5, Figure 6 and Figure 7 show the influence of air mass flow on drying, comparing moisture ratio for the different operating modes of the dryer: mixed passive, mixed active ($q1$ and $q2$), and hybrid active ($q1$ and $q2$).

On the other hand, although the former figures have shown moisture ratio and temperature evolution for the products in the lower tray of the dryer, the results obtained for the assays with two trays are also shown, with the aim of observing how the increase in relative humidity in the air (after going through the first tray) reduced its drying capacity when going through the second one.

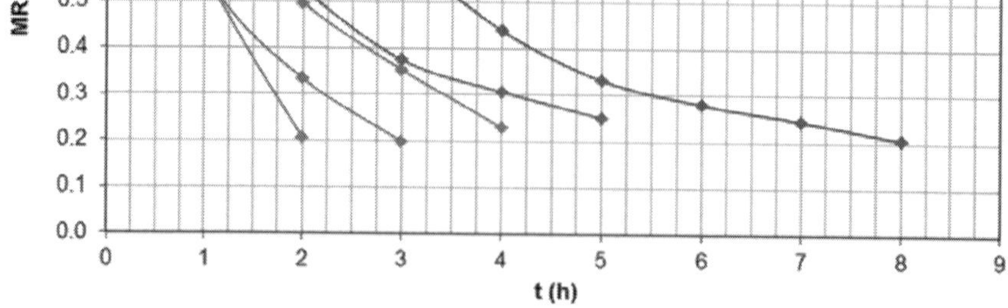

Figure 5. Influence of air mass flow on olive pomace (OP) drying.

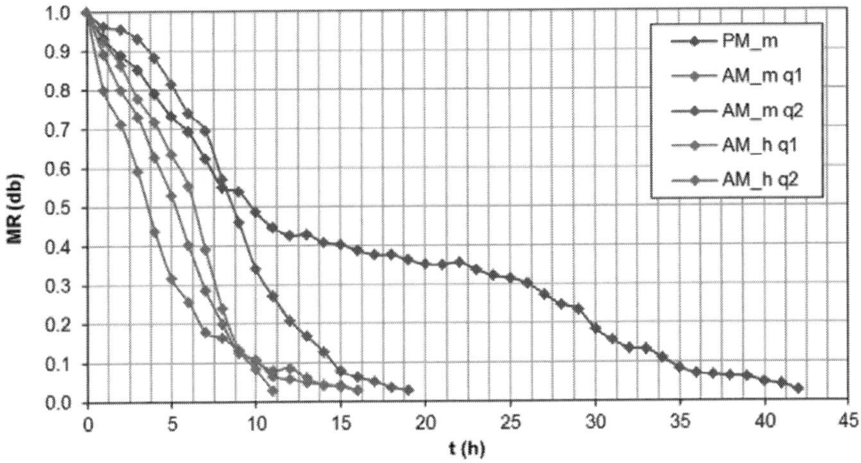

Figure 6. Influence of air mass flow on olive mill wastewater (OMW) drying.

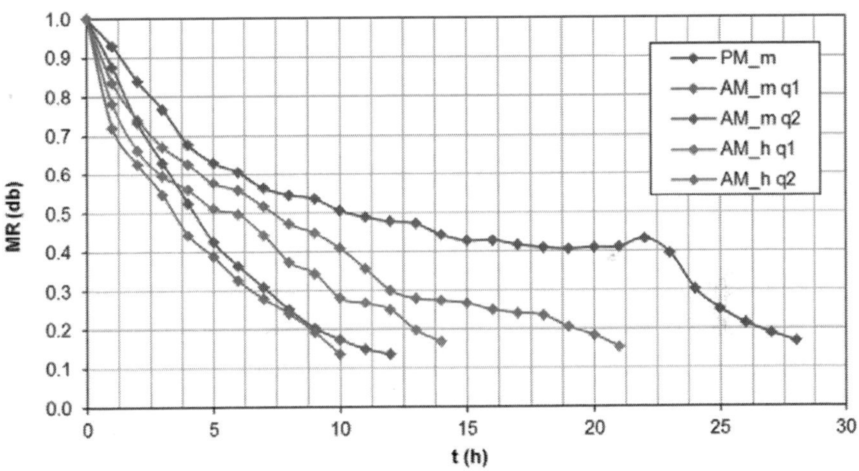

Figure 7. Influence of air mass flow on sludge residue (SLG) drying.

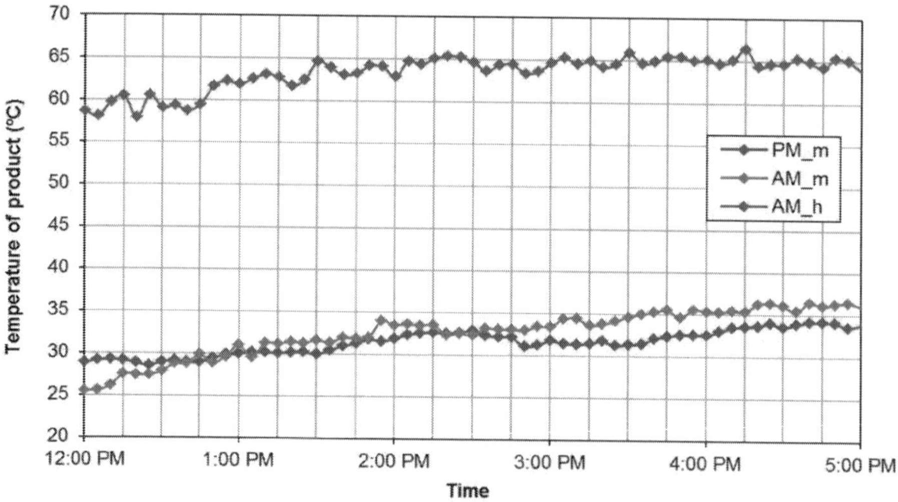

Figure 8. Temperature variation for OP.

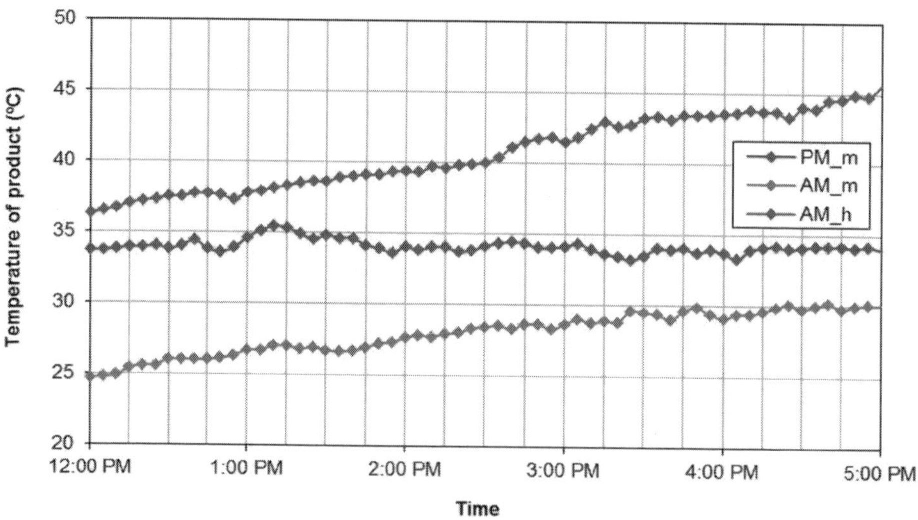

Figure 9. Temperature variation for OMW.

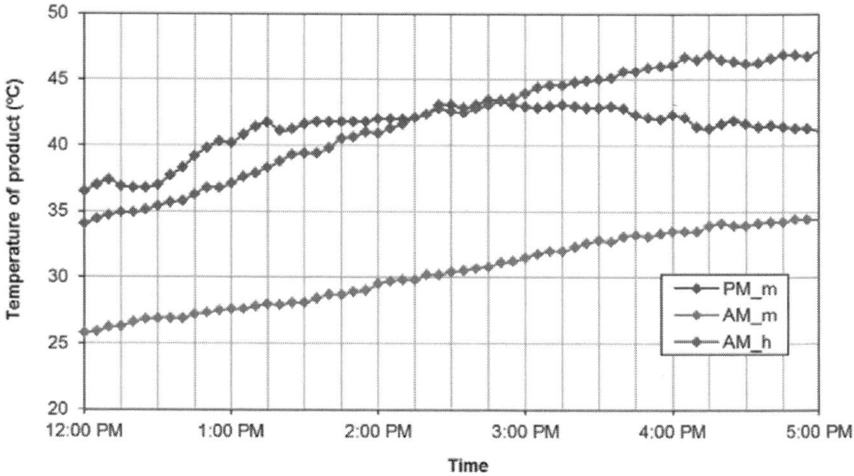

Figure 10. Temperature variation for SLG.

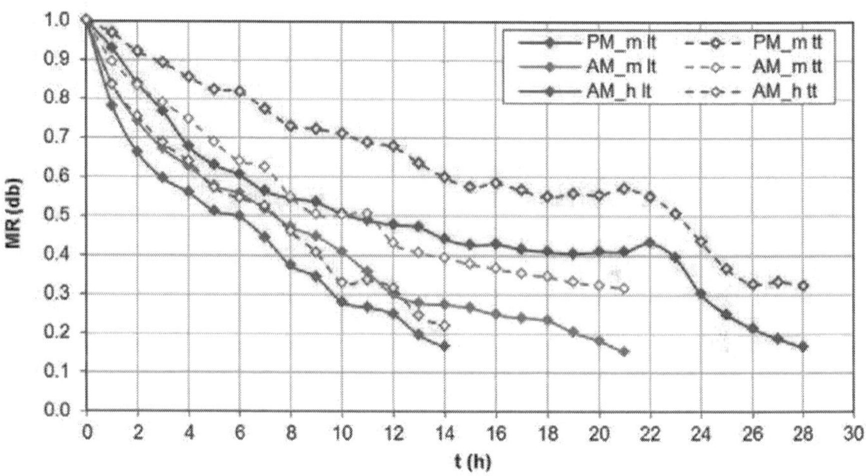

Figure 11. *MR* (db) for different OP trays.

Figure 11, Figure 12 and Figure 13 show moisture ratio for the wastes depending on their location in the dryer (lower tray,*lt*, and top tray, *tt*). For 20% final moisture (wb), fixed as the limit for a product on the lower tray, the corresponding final moisture for the same waste on the top tray was considerably higher, for all the wastes and operating modes, in a range between 15%–50%. In turn, drying times increased (two to three times longer) on the top tray. Similar results were obtained by several authors for grape, fig, pea, tomato and onion drying [31] or rice [32].

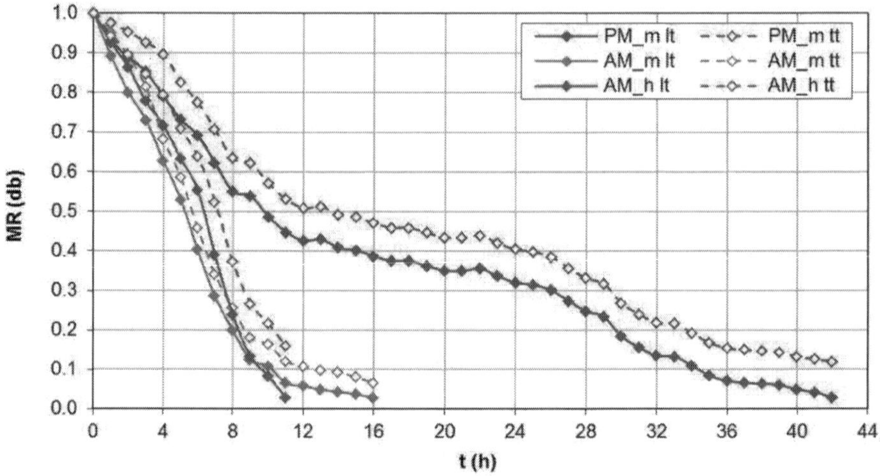

Figure 12. *MR* (db) for different OMW trays.

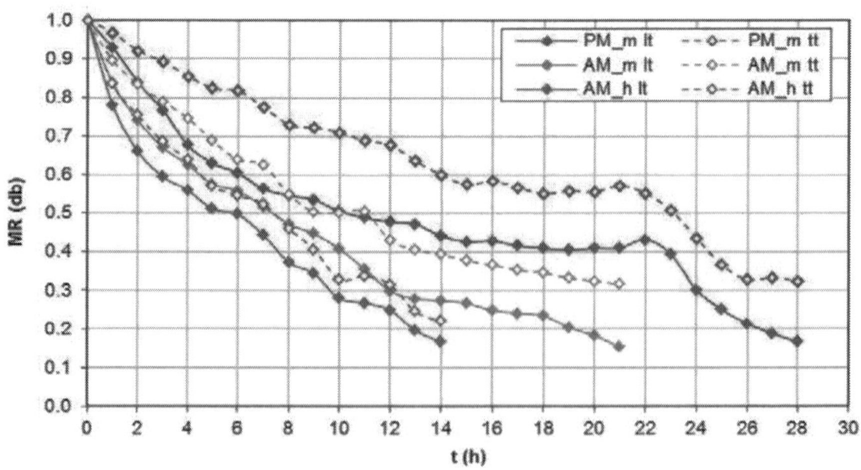

Figure 13. *MR* (db) for different SLG trays.

In all cases, longer drying times were obtained in passive mode. Depending on the operating mode, the drying times were around 2–8 h for OP, 11–42 h for OMW, and 10–28 h for SLG.

In mixed and hybrid active modes, air mass flow was above the optimum value ($q1$), not implying a substantial improvement of drying times, because of air temperature reduction in the chamber, decreasing its drying effectiveness [24,29]. In fact, drying times for OP and OMW with higher air mass flow were longer than the initially obtained ones. On the contrary, an increase in mass air flow contributed to faster drying in the case of SLG.

This circumstance was also observed in hybrid active mode, although air temperature in the collector rose due to the use of resistors. SLG drying was, again, shorter as air flow increased.

Therefore, and with the different drying behavior that was observed, a suitable and accurate adjustment of air mass flow for each waste is vital in order to select the optimum solar drying system [26]. This fact pointed out the importance of temperature and air mass flow control in solar drying systems, depending on the waste to be dried [18].

Another important parameter for drying kinetics is product temperature. If the moisture ratio is known for several temperature values, simulation models can estimate product temperature from drying kinetics or vice versa. Moreover, through this temperature value it is possible to obtain basic parameters in drying evolution of materials, such as convective rates.

Figure 8, Figure 9 and Figure 10 show product temperature changes, measured during the assays of the different drying modes: mixed passive, mixed active, and hybrid active modes, the two latter with $q1$ mass flow. For all the by-products and modes, average air temperature values at the entrance to the chamber were 46, 37 and 57 °C, respectively.

For OP waste, product temperature was similar to those found in mixed passive and mixed active modes, at around 35 °C for natural and forced convection, respectively (due to the use of holed trays in this case). However, in OMW and SLG's case, the temperature found for passive mode was higher than for active mode, being about 35 and 30 °C for OMW and 43 and 35 °C for SLG, respectively. As expected, the highest product temperature values were obtained for hybrid active mode, with approximately 65 °C for OP, 50 °C for SLG, and 45 °C for OMW. In this way, in this kind of dryers, the use of an auxiliary energy source is important, in order to reduce the disadvantages related to direct solar drying (such as climatological changes, difficulty at controlling the process, long drying times, *etc.*) [30], obtaining, in many cases, over 50% reduction in drying times.

Concerning product temperature, the recorded values on the lower tray were higher than the ones obtained on the top tray in all cases, within a range of 2%–10%.

3.3. Influence of External Parameters

The increase in drying air temperature and its relative humidity decrease cause considerable reductions in drying times [23]. Moreover, temperature and humidity depend on environment variables. In order to analyze moisture ratio reduction concerning temperature and relative humidity variations in drying air, two scenarios (*a* and *b*, with *a* being the most favorable one for drying) were studied within the drying chamber, using the average values for these variables that were shown in Table 4. Defining Section 1 and Section 2 as the outside and inside of the drying chamber, respectively, Figure 14,Figure 15 and Figure 16 show the *MR* evolution.

Thus, for the above-mentioned temperature increase and relative humidity decrease (comparing situation *a* and *b*), and 0.3 as *MR* (db) value, 20%, 40% and 35% drying time reductions were obtained for OP, OMW and SLG, respectively.

Table 4. Average values.

Situation	Variables	Wastes		
		OP	OMW	SLG
a	*M2a* (% wb)	10	15	15
	T2a (°C)	45	40	40
b	*M2b* (% wb)	15	25	20
	T2b (°C)	40	30	35

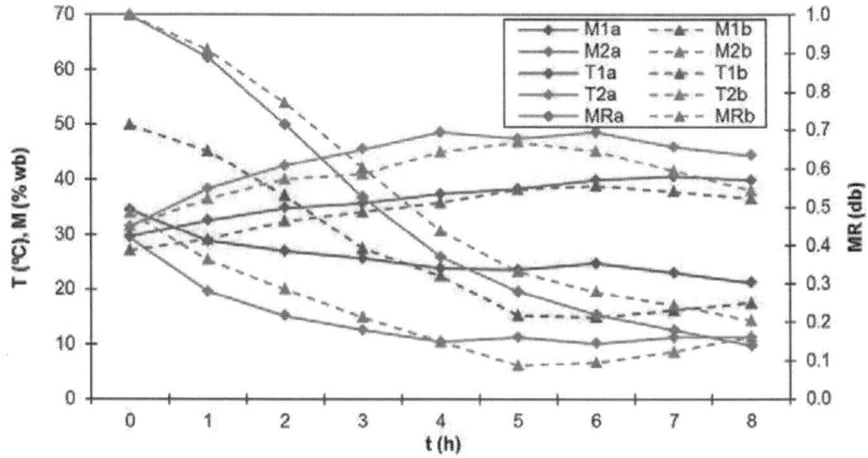

Figure 14. *T*, *M* and *MR* variation of OP for two situations.

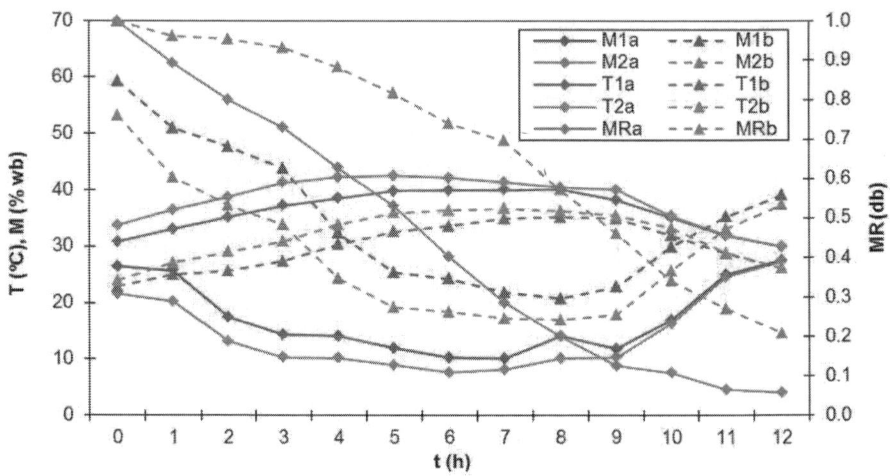

Figure 15. *T*, *M* and *MR* variation of OMW for two situations.

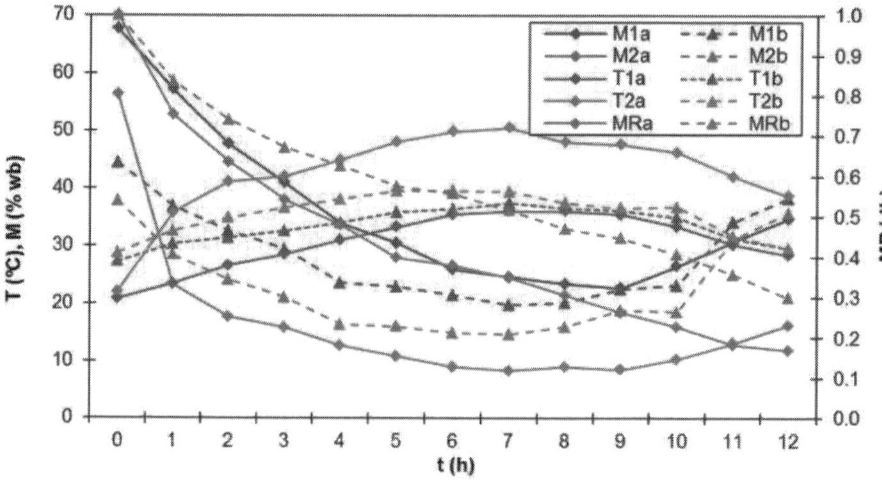

Figure 16. *T*, *M* and *MR* variation of SLG for two situations.

Figure 17 compares product temperature for indirect and mixed active modes, with $q1$ air mass flow (0.22 kg/s). In general, product temperatures were higher in mixed active mode for average radiation between 600 and 900 W/m^2, recording the highest values for OP (35–40 °C), followed by SLG (30–35 °C), and OMW (25–30 °C).

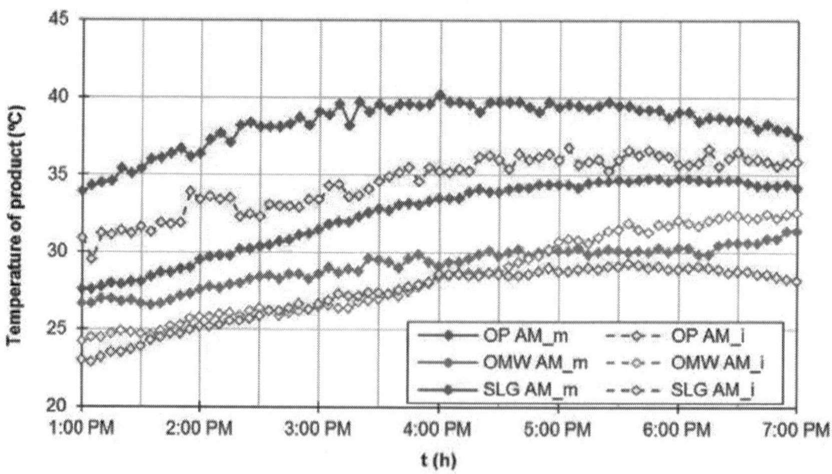

Figure 17. Product temperatures in AM (indirect and mixed types).

3.4. Drying Time and Efficiency of the Drying Chamber

Dryer efficiency assesses the overall thermal performance of the system, being a measurement for the ratio of used energy for drying.

Table 5 shows the values for mixed passive, mixed active, and hybrid active, considering solar radiation contribution, energy consumption for the fan at $q1$ regime and resistors during the drying process.

Table 5. Drying efficiency (%).

Mode	OP	OMW	SLG
Mixed passive	15.3	14.1	13.2
Mixed active	23.2	23.1	15.9
Hybrid active	44.0	26.5	17.3

Drying efficiency was between 13% and 44%, analogous to those values obtained in similar dryers [30,33]. The highest values for this parameter were obtained for hybrid active mode, followed by mixed active and passive modes. In this way, a qualitative improvement in efficiency for active mode compared to passive mode was observed, higher in OP and OMW (>50% and >60%, respectively) than in SLG (20%). In turn, it can be observed that drying efficiency for OP, for each operating mode, was higher than the levels for OMW and SLG.

Lower efficiency values were measured for SLG, not observing significant changes in drying efficiency for the different operating modes.

Finally, Table 6 sums up drying time to get 20% moisture levels (wb), for the most relevant modes and types used in this research work.

In all cases, the use of a mechanical element to contribute to forced convection (for optimum flow), along with the use of an auxiliary energy system, implied the consecution of shorter drying times. The shorter drying times were obtained for OP waste. Drying times for SLG were acceptable and similar to those obtained with other materials in solar dryers. The highest reduction in drying time (from passive mode to active mode) was observed in OMW samples, being a relevant aspect due to the fact that this is considered the most problematic waste in olive oil industry from an environmental point of view.

Table 6. Drying time (h).

Mode	Type	Wastes		
		OP	OMW	SLG
Passive	Mixed	8	42	28
Active ($q1$)	Indirect	7	45	28
	Mixed	4	16	21
	Hybrid	2	11	14

4. CONCLUSIONS

The most relevant findings in the present work were the following:

There was a great influence of solar radiation and air mass flow on exhaust air temperature and collector efficiency. Moreover, there was a limit flow value from which dryer efficiency was not improved.

The longest drying times were obtained in passive mode, being reduced from active mode between 30% and 80%.

Concerning product temperature, the highest values were obtained in hybrid mode, pointing out the importance of using an auxiliary energy source during low solar radiation periods or nighttime.

The increase in air temperature, as well as its relative humidity reduction, caused considerable reductions in drying times, between 20% and 40%, providing suitable values for solar drying use.

The results obtained for OP and OMW wastes, in mixed type, were promising for their solar drying use.

In the case of SLG waste, good drying times were obtained. However, its lower efficiency levels recommend, due to its physical characteristics, the use of turners in order to improve the process and heat transfer and, subsequently, increase solar drying efficiency.

Therefore, this research work shows an alternative to OMW removal-use (a highly pollutant waste). The study that has been carried out enables further knowledge about drying kinetics of olive oil wastes and the results could be extrapolated to a bigger plant, even for other wastes with similar characteristics. For future studies, environmental and economic aspects related to solar drying will be pointed out, as well as some details concerning its implementation on a large scale.

ACKNOWLEDGMENTS

We would like to appreciate the "Gobierno de Extremadura" for the support to research groups (GR10151).

AUTHOR CONTRIBUTIONS

Regarding the contribution of each author to the article, the original idea and work coordination were done by Irene Montero and María Teresa Miranda. The analysis of data was carried out by Francisco José Sepúlveda, José Ignacio Arranz, Carmen Victoria Rojas and Sergio Nogales. Finally, the wording of the conclusions was made by Francisco José Sepúlveda.

GLOSSARY

A_c	Area of the collector (m^2)	$q1, q2$	Air mass flow (kg/s)
BD	Bulk density (kg/m^3 wb)	*T*	Temperature (°C)
db	Dry basis	*tt*	Top tray
FC	Fixed carbon (% db)	*T*	Time (h)
I_t	Solar radiation at time *t* (kJ/m^2)	*VM*	Volatile matter (% db)
HHV	High heating value (MJ/kg db)	W_O	Mass of the product at initial time (kg)
L	Latent heat of vaporization (kJ/kg)	W_t	Mass of the product at time t (kg)
LHV	Low heating value (MJ/kg db)	wb	Wet basis
lt	Lower tray	*AM_h*	Hybrid active mode

M	Moisture (%)	*AM_i*	Indirect active mode
M_0	Initial moisture (%)	*AM_m*	Mixed active mode
M_t	Moisture at time t (%)	*PM_h*	Hybrid passive mode
m_c	Mass of the fuel (kg)	*PM_i*	Indirect passive mode
m_s	Dry mass (kg)	*PM_m*	Mixed passive mode
m_w	Wet mass (kg)	OMW	Olive mill wastewater
MR	Moisture ratio	OP	Olive pomace
E_a	Energy from auxiliary power system (kJ)	SLG	Sludge residue
E_f	Energy from fan (kJ)		

REFERENCES

1. Dermeche, S.; Nadour, M.; Larroche, C.; Moulti-Mati, F.; Michaud, P. Olive mill wastes: Biochemical characterizations and valorization strategies. *Process Biochem.* **2013**, *48*, 1532–1552.
2. Parejo, J.M.; Rubio, J.C.; Pérez, S.; Soriano, M. Health and safety management in olive oil mills in Spain. *Saf. Sci.* **2013**,*51*, 101–108.
3. Peri, C.; Proietti, P. Olive mill waste and by-products. In *The Extra-Virgin Olive Oil Handbook*; John Wiley & Sons, Ltd.: Chichester, UK, 2014; pp. 283–302.
4. García-Maraver, A.; Zamorano, M.; Ramos, A.; Díaz, L.F. Analysis of olive grove residual biomass potential for electric and thermal energy generation in Andalusia (Spain). *Renew. Sustain. Energy Rev.* **2012**, *16*, 745–751.
5. Gómez, F.J.; Casanova, P.J.; Palomar, J.M.; Cruz, F. Modeling of olive-oil mill waste rotary dryers: Green energy recovery systems. *Appl. Therm. Eng.* **2015**, *80*, 362–373.

6. Miranda, M.T. Thermal Utilization of Biomass. Application to SLG. Ph.D. Thesis, University of Extremadura, Badajoz, Spain, 10 June 2005. (In Spanish).
7. Gigliotti, G.; Proietti, P.; Said-Pullicino, D.; Nasini, L.; Pezzolla, D.; Rosati, L.; Porceddu, P.R. Co-composting of olive husks with high moisture contents: Organic matter dynamics and compost quality. *Int. Biodeterior. Biodegrad.* **2012**, *67*, 8–14.
8. Nasini, L.; Gigliotti, G.; Balduccini, M.A.; Federici, E.; Cenci, G.; Proietti, P. Effect of solid olive-mill waste amendment on soil fertility and olive (*Olea europaea* L.) tree activity. *Agric. Ecosyst. Environ.* **2013**, *164*, 292–297.
9. Meziane, S. Drying kinetics of olive pomace in a fluidized bed dryer. *Energy Convers. Manag.* **2011**, *52*, 1644–1649.
10. Vega-Gálvez, A.; Miranda, M.; Díaz, L.P.; López, L.; Rodríguez, K.; Scala, K.D. Effective moisture diffusivity determination and mathematical modelling of the drying curves of the olive-waste cake. *Bioresour. Technol.* **2010**, *101*, 7265–7270.
11. Milczarek, R.R.; Dai, A.A.; Otoni, C.G.; McHugh, T.H. Effect of shrinkage on isothermal drying behavior of 2-phase olive mill waste. *J. Food Eng.* **2011**, *103*, 434–441.
12. Montero, I. Modeling and Construction of Hybrid Solar Dryer for Biomass by Products. Ph.D. Thesis, University of Extremadura, Badajoz, Spain, 30 November 2005. (In Spanish).
13. VijayaVenkataRaman, S.; Iniyan, S.; Goic, R. A review of solar drying technologies. *Renew. Sustain. Energy Rev.* **2012**,*16*, 2652–2670.
14. Sharma, A.; Chen, C.R.; vu Lan, N. Solar-energy drying systems: A review. *Renew. Sustain. Energy Rev.* **2009**, *13*, 1185–1210.
15. Fudholi, A.; Sopian, K.; Ruslan, M.H.; Alghoul, M.A.; Sulaiman, M.Y. Review of solar dryers for agricultural and marine products. *Renew. Sustain. Energy Rev.* **2010**, *14*, 1–30.
16. Miranda, T.; Arranz, J.I.; Montero, I.; Román, S.; Rojas, C.V.; Nogales, S. Characterization and combustion of olive pomace and forest residue pellets. *Fuel Process. Technol.* **2012**, *103*, 91–96.
17. Montero, I.; Miranda, M.T.; Arranz, J.I.; Rojas, C.V. Thin layer drying kinetics of by-products from olive oil processing. *Int. J. Mollecular Sci.* **2011**, *12*, 7885–7897.

18. Montero, I.; Blanco, J.; Miranda, M.T.; Rojas, S.; Celma, A.R. Design, construction and performance testing of a solar dryer for agroindustrial by-products. *Energy Convers. Manag.* **2010**, *51*, 1510–1521.
19. Augustus Leon, M.; Kumar, S.; Bjhattacharya, S.C. A comprehensive procedure for performance evaluation of solar food dryers. *Renew. Sustain. Energy Rev.* **2002**, *6*, 367–393.
20. El-Sebaii, A.A.; Aboul-Enein, S.; Ramadan, M.R.I.; El-Gohary, H.G. Experimental investigation of an indirect type natural convection solar dryer. *Energy Convers. Manag.* **2002**, *43*, 2251–2266.
21. Liu, Z.; Quek, A.; Balasubramanian, R. Preparation and characterization of fuel pellets from woody biomass, agro-residues and their corresponding hydrochars. *Appl. Energy* **2014**, *113*, 1315–1322.
22. De la Gómez Cruz, F.J.; Palomar, J.M.; Casanova, P.J.; Cruz, F. Experimental determination of effective moisture diffusivity during the drying of clean olive stone: Dependence of temperature, moisture content and sample thickness. *Fuel Process. Technol.* **2015**, *137*, 320–326.
23. Doymaz, I. Drying behaviour of green beans. *J. Food Eng.* **2005**, *69*, 161–165.
24. Babalis, S.J.; Belessiotis, V.G. Influence of the drying conditions on the drying constants and moisture diffusivity during the thin-layer drying of figs. *J. Food Eng.* **2004**, *65*, 449–458.
25. Velic, D.; Planinic, M.; Tomas, S.; Bilic, M. Influence of airflow velocity on kinetics of convection apple drying. *J. Food Eng.* **2004**, *64*, 97–102.
26. Corvalan, R.; Horn, M.; Roman, R.; Saravia, L. *Engineering Solar Drying*; Science and Technology Program for Development of V Centenary (CYTED-D): Madrid, Spain, 1992.
27. Simate, I. Optimization of mixed-mode and indirect-mode natural convection solar dryers. *Renew. Energy* **2003**, *28*, 435–453.
28. Gögüs, F.; Maskan, M. Air drying characteristics of solid waste (pomace) of olive oil processing. *J. Food Eng.* **2006**, *72*, 378–382.
29. Madamba, P.S.; Driscoll, R.H.; Buckle, K.A. The thin-layer drying characteristics of garlic slices. *J. Food Eng.* **1996**, *29*, 75–97.
30. Bena, B.; Fuller, R.J. Natural convection solar dryer with biomass back-up heater. *Solar Energy* **2002**, *72*, 75–83.

31. El-Sebaii, A.A.; Aboul-Enein, S.; Ramadan, M.R.I.; El-Gohary, H.G. Empirical correlations for drying kinetics of some fruits and vegetables. *Energy* **2002**, *27*, 845–859.
32. Jain, D. Modeling the system performance of multi-tray crop drying using an inclined multi-pass solar air heater with in-built thermal storage. *J. Food Eng.* **2005**, *71*, 44–54.
33. Nijmenh, M.N.; Ragab, A.S.; Emeish, M.S.; Jubran, B.A. Design and testing of solar dryers for processing food wastes.*Appl. Therm. Eng.* **1998**, *18*, 1337–1346.

3
CHAPTER

TOWARD IMPLEMENTING HH THE AMIR DECLARATION OF 2% ELECTRICITY GENERATION BY SOLAR ENERGY IN 2020

Mohamed Darwish, Ashraf Hassan, Rabi Mohtar

Qatar Environment and Energy Research Institute, Doha, Qatar

ABSTRACT

The utility solar power plants were reviewed and classified by two basic groups: direct thermal concentrating solar power (CSP) and photovoltaic (PV). CSP as Parabolic Trough Collector (PTC) of 100 MW solar power plants (SPP) is suggested and suitable to provide solar thermal power for Qatar. Although, LFC had enough experience for small projects, it is still need to work in large scale plant such as 100 MW and couple with multi effect distillation (MED) to confirm costs.

Keywords: Concentrated Solar Power; Linear Fresnel Collector; Desalination

1. INTRODUCTION

HH the Amir of Qatar declared at the end of COP 18 in Dec. 2012 that: By 2020, solar energy should generate at least 2% of Electric Power (EP) produced in the country.

Qatar's Power Plants (PPs) capacity in 2010 was around 8000 MW, and this is expected to be doubled in 2020, or becomes 16,000 MW. The PPs capacity was doubled in only six years in the last decade, as shown in **Figure 1**, [1]. The main operating PPs types in Qatar are Gas Turbine (GT), and GT Combined Cycle (CC). These plants have much higher Capacity Factor (CF) than Solar Power Plants, see **Figure 2**, [2]. The annual CF of the Qatar's PPs was more than 40% in 2010, because of the winter EP low load. The CF of the Solar Power Plants (SPPs) is in the range of 20%. This means that, by 2020, the total capacity of the Solar Power plants (SPPs) should have at least 640 MW in order to generate 2% of the EP. The range of CF shown in **Figure 2** is when the plants are operating as base load plants.

Solar Energy (SE) is as an attractive, available, clean, and free primary energy source. Application of SE to generate EP helps in growing new technologies and enhances energy security. Although the SPPs' are more expensive than the presently used GTCC, it is logic that Qatar supports solar energy development as this is likely to be one of the main EP generation methods in the future. Concentrated Solar Power (CSP) plants offer dispatchable power when integrated with Thermal Energy Storage (TES). The EP outputs of the SPPs match very well with the peak load demand due to air conditioning loads during summer, as this depends on solar insolation along the day. The SPPs are increasingly moving into the range which has traditionally been the domain of classic large-scale PPs. The existing trend of falling SPPs costs would accelerate adoption of SE and reduce the need for incentives. Other advantage of using SPPs is the deep cut in the greenhouse gases emission, mainly caused by fossil fuel combustions in conventional PP.Solar collector manufacturing is conventional technology, can be manufactured locally with dropping cost as the need increases, and creates new business and jobs.

Domination of NG as primary fuel in Qatar can solve the main SE problem of intermittent nature, by using hybrid solar-NG plants. Compared to PPs using NG, the SPPs are in their infancy, and are expected to become less expensive as technologies mature and become mass produced.

The major barrier for the technology at present is the cost of generated electricity by SE. Meanwhile, continuous increase of the NG cost, **Figure 3**, [3], is another reason for looking forward to diversify the primary energy for EP generation, and use of SE.

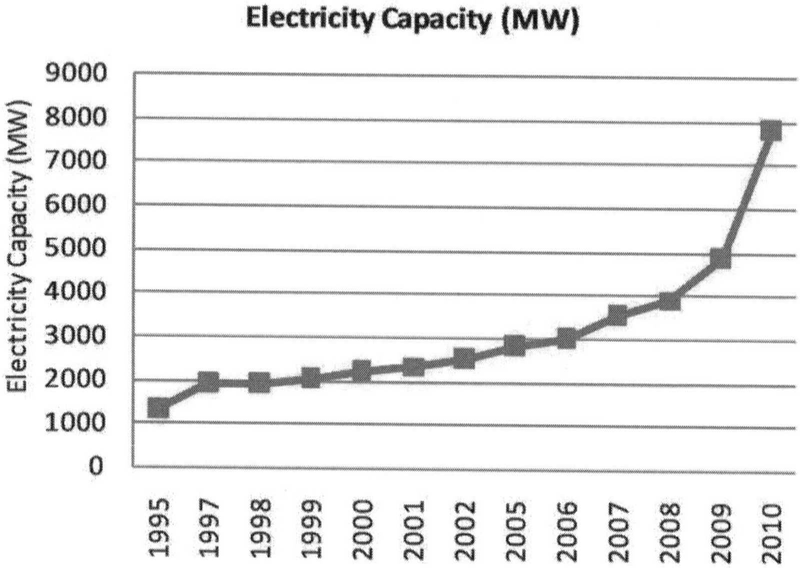

Figure 1. History of the EP installed capacity, [1].

2. UTILITY SOLAR POWER PLANTS (SPP) BROAD CLASSIFICATIONS

The utility-scale SPPs are generally classified by two basic groups: thermal concentrating solar power (CSP) and photovoltaic (PV). These are the types to be considered for the new SPPs in Qatar. Since the new required capacity is in the range of 640 MW, it is logic to choose plants of 50 MW - 100 MW to avoid the use too many small units.

2.1. Thermal Concentrated Solar Power (CSP) Plants

In thermal Concentrated Solar Power (CSP), solar receivers (mirrors) reflect their incident direct solar rays as concentrated rays on receivers, where Heat Transfer Fluids (HTF) is flowing. The HTF is heated to become heat source for driving heat engines, such as steam turbines, gas turbines or Stirling engines, producing EP, **Figure 4**. So, in CSP plants, only direct radiation is utilized, while diffuse type is not. The most developed and used type of concentrated solar collectors is the Parabolic Trough

(PTC), followed by power tower. Other types are the Linear Fresnel (LFC), and parabolic dish operating Stirling engine are under development to reach commercial status at large capacities.

Another mode of SPPs' operation is the use of thermal storage. Solar heat collected during the daytime can be stored in liquid or solid media such as molten salts, ceramics, concrete or phase-change media. The stored heat can then be extracted at night for running the power block; and thus the plant becomes dispatchable. The PP's dispatchability is very crucial in PPs as the plant can respond by generating the required power as the load demand. It can be turned on or off, or can adjust their power output on demand.

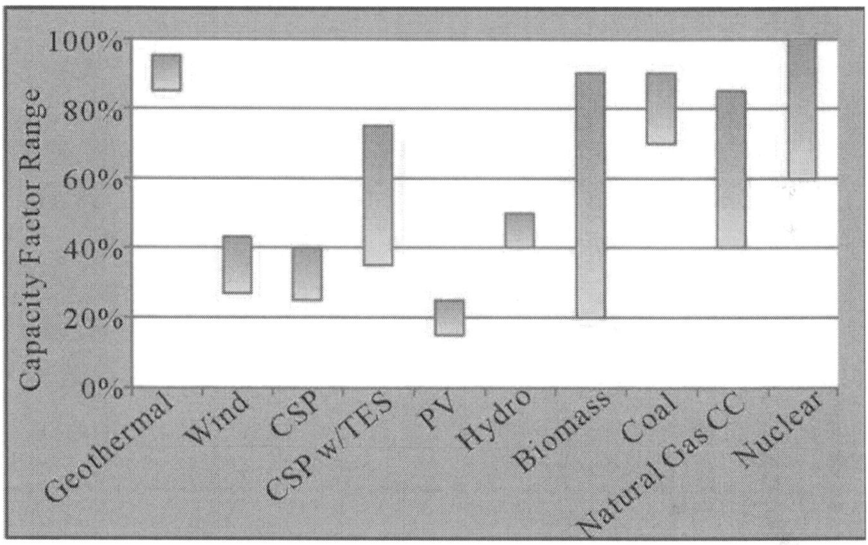

Figure 2. Comparison of capacity factor by technologies, [2].

The capacity of individual CSP plants is now typically between 50 and 280 MW, but could still be larger.

2.1.1. Concentrated Solar Power (CSP) Using Parabolic Trough Collectors (PTC)

The PTCs are the most used type of collectors in SPPs. The PTCs have mirrors of linear parabolic shape, reflect concentrated rays onto tube-shape receiver positioned along the reflector's focal line, and carrying the HTF, **Figure 5**, [5]. More on the PTC plants are given later. The components of the SPP using PTC and molten salt thermal energy storage are shown in **Figure 6**, [6].

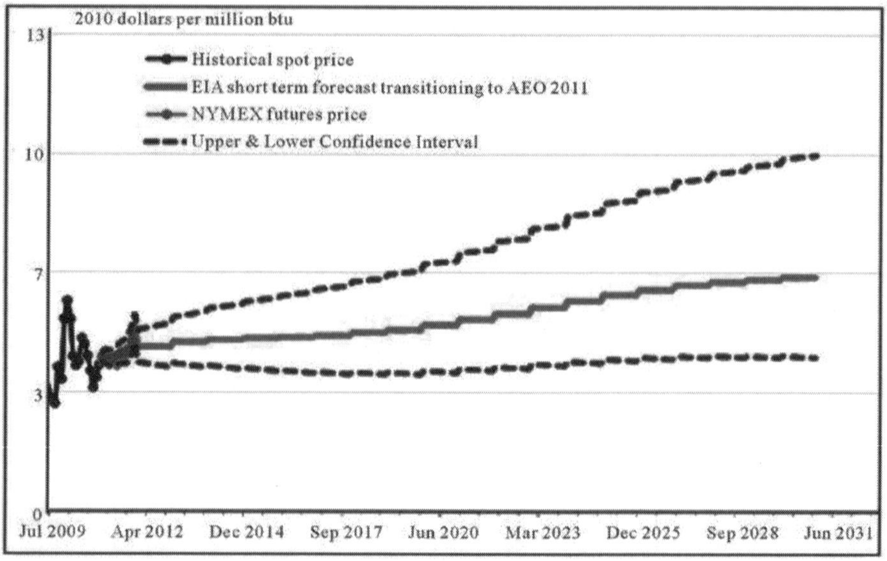

Figure 3. Henry hub wholesale natural gas price history, [3].

2.1.2. Concentrated Solar Power (CSP) Using Linear Fresnel Collectors (LFC)

The LFC concept uses sets of long mirror facets that reflect light to a linear receiver where it can be directly absorbed to generate steam. The LFCs are much cheaper and need much lighter carrying structure to mount compared to the PTCs. The optical performance of the LFC system is limited by the angle at which the sunlight strikes the reflectors, and because the mirrors must be oriented to reflect the irradiation to the receiver, they most often do not directly face the sun. This non-normal orientation towards the incoming sunlight is the primary optical loss for LFC, and the loss is incurred both with respect to the transversal plane (perpendicular to the axis of the collector) and longitudinal plane (parallel with the axis of the collector). This concept is illustrated in Figures 7(a) and (b), [7]. **Figure 7**(c) shows the arrangement of LFC in SPP. There is growing interest in Linear Fresnel Collectors (LFC) technology as candidate for combined-cycle integration with solar energy through direct steam generation (DSG).

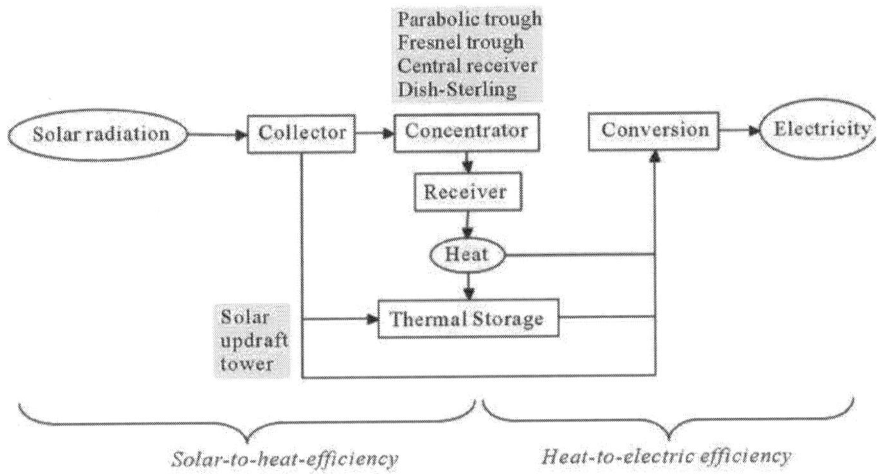

Figure 4. Schematic illustration of thermal CSP plants, [4].

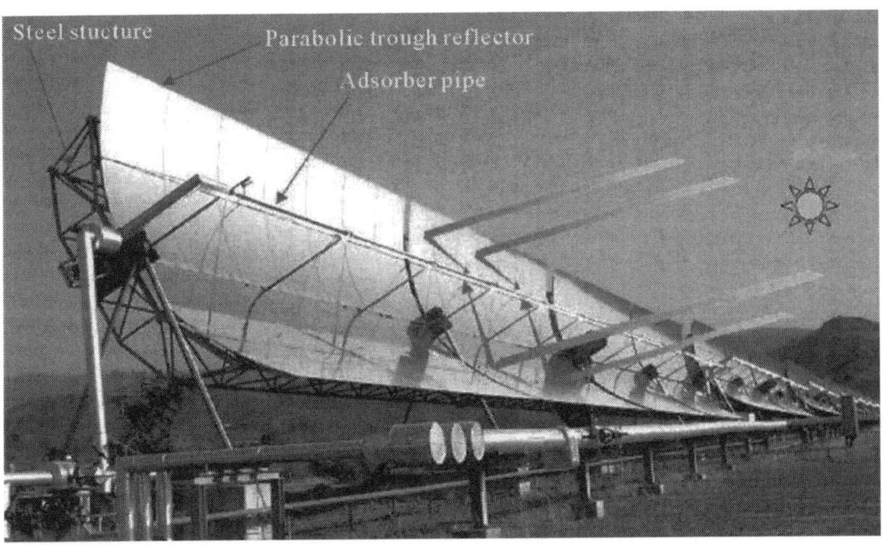

Figure 5. Photo of parabolic trough system (source: NREL), [5].

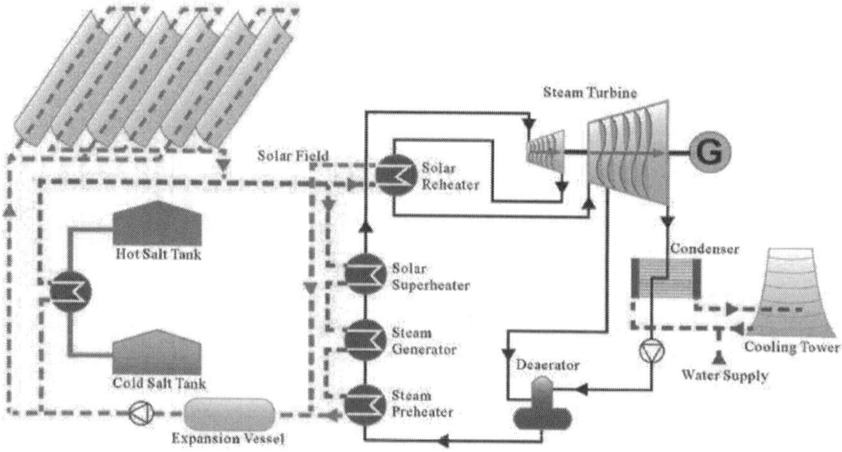

Figure 6. SPP arrangement using rankine steam cycle with thermal energy storage, [6].

The first large utility-scale basis of 30 MW using Linear Fresnel Collectors (LFC) to generate steam operating power cycle was Puerto Errado 2 (PE2), [8]. The plant

covers an area of 700,000 m^2 land with 28 rows of mirrors, and has aperture area of 302,000 m^2, almost 10,000 m^2/MW. The ratio of land to solar collector area is 2.33. This land to apertures' area of 2.33 is low when compared with 4 for PTC plants. This plant, operated by Novatec Solar España, is a linear Fresnel reflector system, **Figure 8**, [8].

2.1.3. Concentrated Solar Power (CSP) Using Power Tower

A solar power tower consists of an array of distributed dual-axis tracking flat mirrors (called heliostats) that concentrate solar rays on a central receiver atop a tower; and containing the HTF, see **Figure 9**, [9].

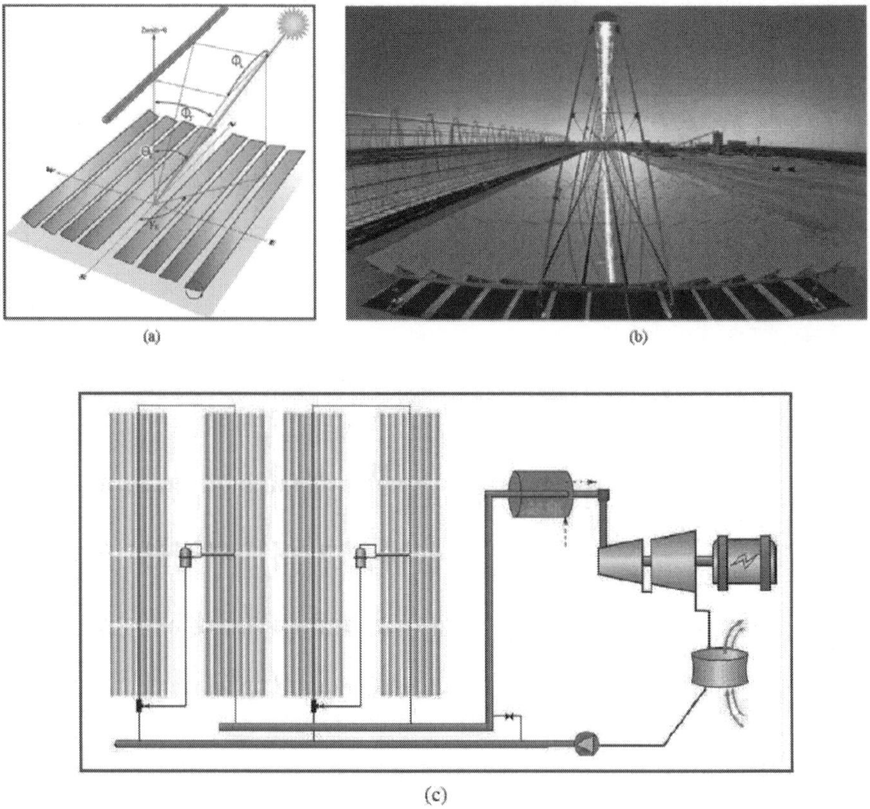

Figure 7. (a) Angles associated with the optical performance of the LFR technology, including transversal incidence φT, longitudinal incidence φL, solar zenith θz, and solar azimuth γs, [7]. (b) The linear fresnel collector type of direct steam generator, [6]. (c) SPP Arrangement using FLC and rankine steam cycle without thermal energy storage, [7].

The HTF heated up to 500°C - 1000°C and then is used as a heat source for a power heat engine or energy storage system. Power tower with molten-salt thermal storage was developed specifically to generate EP. It can be the most efficient and lowest cost solar power systems but it is not commercially established yet.

The molten-salt provides efficient, low-cost thermal energy storage (TES) system, and allows solar plants to be designed with high annual Capacity Factor (CP) or being dispatch-able to meet summer load.

Figure 8. An overview of the errado 2 plant, [8].

The PS20 is the world's most powerful solar power tower of 20 MW. The PS20 was designed by Abengoa Solar in Seville, Spain. Its operation began in April 2009. It has 162 m solar tower height and 1255 heliostats. The PS20 is adjacent to the world's first commercial CSP tower, the PS10, also designed by Abengoa Solar, [10], see **Figure 10**.

2.1.4. Concentrated Solar Power (CSP) Using Parabolic Dish

Parabolic dish systems use mirrors forming dish-shaped reflector to focus solar rays onto receiver at the dish focal point, **Figure 11**, [11]. A working fluid, heated in the receiver, such as hydrogen drives a turbine or Stirling heat engine. Most current dish applications use Stirling engine technology because it has high solar-to-electric efficiency (@30%). Its modular nature (25 kWe units) implies that plants of virtually any size could be built or expanded. These systems do not require water for cooling; another benefit compared to PTC and LFC systems. Current systems have not demonstrated the level of reliability considered necessary for large commercial system.

(a)

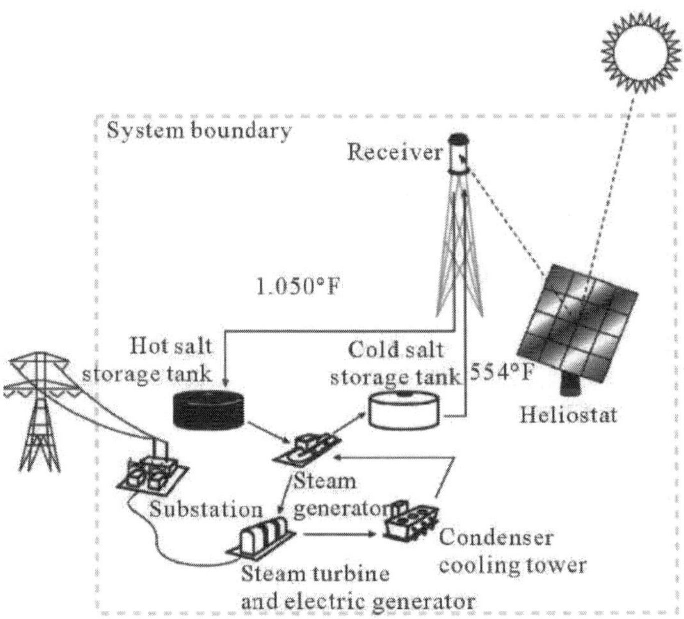

(b)

Figure 9. (a) Figure of solar power tower with heliostat, [9]. (b) Schematic diagram of solar tower power cycle, [9].

Figure 10. Solar power towers for PS10 and PS20, [10].

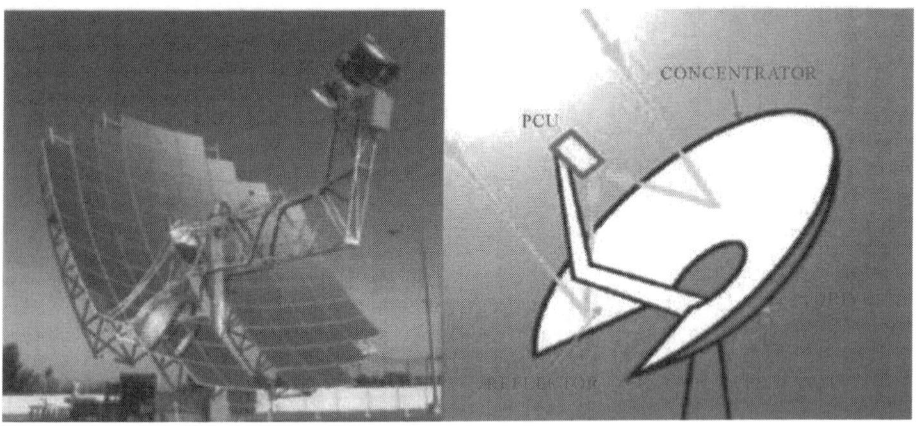

Figure 11. Stirling dish systems at sandia national labs, [11].

At present, there are no operating utility large scale parabolic dish plants.

Currently, there is one dish/stirling plant containing 60 dishes in operation in Arizona, US. Each dish has 250 kW capacities. So, the total capacity of the plant is 1.5 MW. This plant started production on January 2010. There are no operating (or in development) commercial dish systems of large utility scale capacity. The NREL gave the following data on Maricopa Solar Project (Maricopa) plant in the US, [12].

2.2. Photovoltaic (PV) Power Station

A PV panel generates electricity when light energy (photon) hits the semiconductor surface of the panel and releases electrons that would generate voltage differences. Large numbers of PV panels are needed to generate electricity on a large scale. The PV systems efficiency has increased during the last decade and reached a maximum of 20 per cent in some commercial panels. The system uses solar panels consisting of number of solar cells manufactured from semiconductors materials. The semiconductors materials exhibit the PV effect. Materials presently used for PV include crystalline silicon (c-Si) such as mono-crystalline silicon, polycrystalline silicon, and amorphous silicon; and cadmium telluride, and copper indium gallium selenide/sulfide. Masdar (in UAE), has already connected a PV power station 10 MW PV to the grid. Masdar is planning to install the first large solar PV plant of 100 MW (called NOOR 1).

Concentrating photovoltaic (CPV) systems are currently developed by different companies. Similar to dish/ Stirling systems, these systems are modular in nature (25 - 50 kWe units) and have potential for high solar-toelectric efficiency (>30%). These systems also do not require water for cooling. Manufacturers are currently providing CPV systems but only with a few MW per year; they still have limited operational experience.

A photovoltaic (PV) power station is large-scale PV system that generates EP to be supplied to the electric grid at the utility level, and not to local users, see **Figure 12**, [13]. PV systems usually include either crystalline silicon (c-Si) or thin-film technologies. Thin film includes an array of advanced materials, but only one— cadmium telluride (CdTe) has had significant success in utility-scale solar development.

2.3. Integrated Solar Combined Cycle Plants (ISCC)

An Integrated Solar Combined Cycle (ISCC) plant consists of conventional Gas Turbine Combined Cycle (GTCC), or simply (CC), plant integrated with solar field, **Figure 13**. The components of the ISCC are Gas Turbine (GT), Heat Recovery Steam Generator (HRSG) driven by the hot gases exhausted from the GT, steam turbine, and concentrated solar collector field (usually PTC but it can be FLC) producing steam. The steam generated in the solar field augments the steam generated from the HRSG, and thus increasing the size and power output of the steam turbine. The ISCC system gives the benefits of solar energy and the highly efficient GTCC operated by Natural Gas (NG), which is abundant in Qatar. In Integrated Solar Gas Turbine Combined Cycle (ISCC), the energy generated by solar energy represents 10 - 20 percent of the total plant output. Kuwait has completed a technical feasibility study for the construction of ISCC plant having the capacity of 280 MW, including 60 MW solar components.

Figure 12. PV power system of 24 MW: SinAn, South Korea of 600,000 m^2 area, [13].

Choice of SPP for Qatar from the Available SPP Alternatives

The required additional capacity of the PP in Qatar is more than 8 GW from 2010 to 2020, and 640 MW of this capacity should be SPPs to satisfy HH the Amir decision to generate at least 2% of EP by solar energy. A logical capacity choice for each SPP should be in the range of 50 - 100 MW. The already existing PPs capacities in Qatar are: 497 MW for Ras Abu Fontas A, 2273 for totalRas Abu Fontas B, 3550 MW for totalRasLaffan A, B, and C, and 2007 MW for Mesaieed Power station, a total capacity of 7830 MW, [1].

Moreover the typical demand for base-load cheap electricity is usually supplied by large units. Thus, the estimated capacity of any suggested PP is in the range of 50 MW or higher. It is noticed that PV power stations (PVPS) the status of commercial well proven type PP for the range of 50 MW - 100 MW suitable for Qatar, and should be considered. However, all PVPS are not dispatchable and have low capacity factor. This means that stand by station of the same PVPS capacity should be built to count on the PVPS.

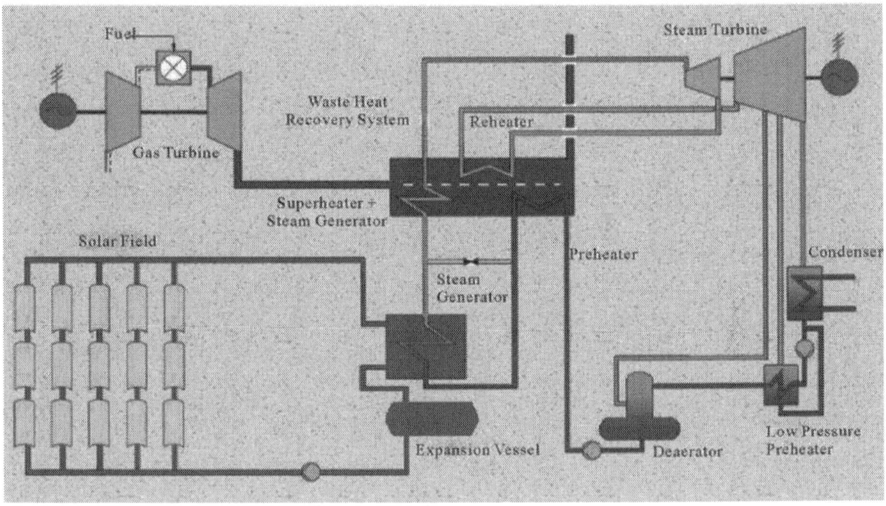

Figure 13. Integrated solar combined cycle ISCC, [14].

Concerning the CSP systems, both the power tower, and the FLC reached the commercial status for capacity of 20 MW, and 30 MW respectively. However the number of these types of plants are so little to give reliable records for their operation and maintenance. These two types are hardly to be considered for capacity of 100 MW for the time being.

2.4. Solar Power Plants (SPP) Using Parabolic Trough Collectors (PTC)

Reviewing the LFC, power tower, and parabolic dish with Stirling engine power stations indicates that none of them reach the commercial status of well proven type with reliable record to build new SPP having capacity of 50 MW or more here in Qatar for the time being. **Table 1**, [6], presents a comparison between the four types of the thermal CSP plants. This table shows that the technical risk of adopting the PTC type is low, while for the risk for power tower and the LFC and power tower is medium, and for the parabolic dish and Stirling engine the risk is high, [6]. The Solar Power Plants (SPP) using Parabolic Trough Collectors (PTC) is the only option left among the CSP type of SPPs. This is the reason to consider only the PTC plants here in details. **Figure 14** shows the installed capacity of the different thermal CSP plants, with clear indication that the SPPs using PTC are the most used and planned type.

The operational large CSP using PTC includes: 354

Table 1. Comparison of solar thermal concentrating technologies, [6].

	Parabolic Trough	Tower	Dish Engine	Linear Fresnel
Commercial experience	➢ 20 years	<4 years	-	-
Technology risk	Low	Medium	High	Medium
Optimal scale/modularity	50 MW to >100 MW	50 MW to >100 MW	100 MW to >100 MW	50 MW to >100 MW
Construction requirement	Demanding	Demanding	Moderate	Simple to moderate
Operating temperature	300°C - 550°C	260°C - 570°C	750°C	270°C
Efficiency	14% - 16%	15% - 22%	24% - 31%	9% - 11%
Storage	Yes	Yes	No	No
Levelized cost of energy ($/kWh)	Current: 0.30 - 0.75 Future: 0.06 - 0.08	Current: 0.20 - 0.90 Future: 0.06 - 0.08	Future: 0.05 - 0.08	Future: 0.06 - 0.08
Water usage	High	High	Low	Medium
Land requirement	High	high	Variable/flexible	Variable
Leading developers	Acciona Solar, Abengoa Solar/Abener, Solar Mellenium, Solel/Siemens	Abengoa Solar/Abener, Bright Source Energy, Torresol, eSolar	Tessera Solar/Stirling Energy Systems (SES)	Ausra/Area (small scale projects), Novatec Solar

MW Solar Energy Generating Systems power installation in the USA, Solnova Solar Power Station (Spain, 150 MW) and Andasol solar power station (Spain, 150 MW).

There is the 370 MW Ivanpah Solar Power Facility, located in California's Mojave Desert, is the world's largest solar thermal power plant project currently under construction. The Solana Generating Station is also under construction and has capacity of 280 MW, and is located 110 km southwest of Phoenix, Arizona.

The SPPs using PTC technology are using steam Rankine cycle as power block. The average capacity is 50 MW or more, which is considered suitable for Qatar. This is the most used SPPs with enough operating experience. Nine SPPs, with a total generation capacity of 354 MW, called SEGS (Solar Electric Generating Systems) were the first large CSP to be built and operated in the California Mojave desert from 1984 - 1990. The SEGS systems are co-fired with natural gas to provide continuous operation when the sun does not shine. These were followed by several SPPs in many part of the world.

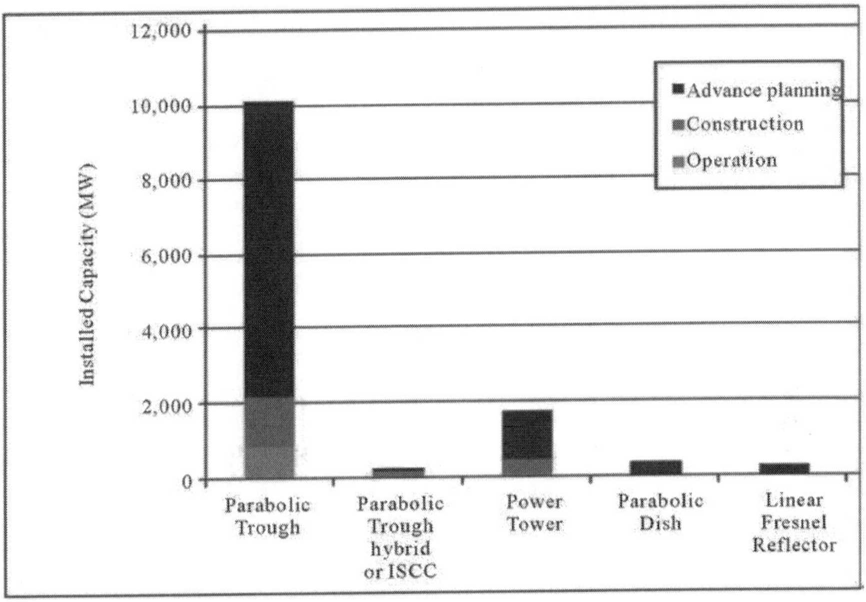

Figure 14. Implementation of CSP technologies, [6].

Table 2. Examples of SPP using PTC, and their basic characteristics, [15].

Name	Capacity MW	Storage hours	DNI kWh/m²/y	Solar field area As/MW	Plant, 1000 m²/MW	¢/kWh	Temperature Solar field in/out	Efficiency
Alvarado1	50		2174		27	27	293/393	
Andasol 1	50	7.5	2136	10202.4	40	27	293/393	16
Andasol 2	50	7.5	2136	10202.4	40	27	293/393	16
La Florida	50	7.5		11055	40	27	298/393	14
Extresol-2	50	7.5	2168	10202.4	40	27	293/393	16
Extresol-1	49.9	7.5	2168	10222.85	40	27	293/393	16
Ibersol Ciudad Real	50	0	2061	5755.2	30		304/391	
La Dehesa	49.9	7.5		11077.15	40		29/393	14
Majadas	50		2142			27		
Manchasol-1	49.9		2208		40		293/393	16

Examples of the several built SPP with PTCs are: 64 MW power plants near Boulder City, US, and several 50 MWe power plants in Spain shown in **Table 2**, [15].

3. SUGGESTED SPP USING PTC

A solar CSP parabolic trough PP of 100 MW net power output is suggested in this study with 6 hours (h) Thermal Energy Storage (TES) system to assure the plant capability to cover the peak load period and to raise the CF to 38%. Tracking the sun from east to west, while rotating on a north-south axis, the trough focuses the sun's energy on a pipe located along its focal line. Troughs can also rotate on an east-west axis only but yield less annual energy. The HTF, typically oil, is heated up to 400°C temperatures while it is circulated through the receiver pipes and then pumped to a central power block area, where it passes through a heat exchanger. In this heat exchanger, called Steam Generator (SG), the oil's heat is then passed to water to generate steam. The steam is used in turn to drive a conventional turbine generator. Since the parasitic power loss of this type of PP is about 10%, the gross power would be 110 MW. The System Advisor Model (SAM) computer program developed by NREL is used here to figure the solar field, the power cycle, the TES, required budget of the project, and finally levelized Energy Cost (LEC), [16]. Because no boundary conditions data are available for Qatar, the environmental data for Cairo, Egypt is used in running the program. This environmental data are given as follows:

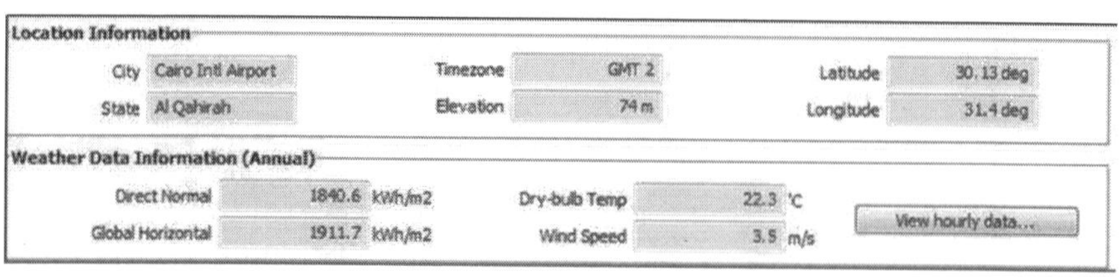

The SPP data are: gross power output of 111 MW, net output to gross output efficiency is 90%, and then the estimated output design is 100 MW. The HTF has temperature to the power block at 391°C, returns at 293°C. The steam throttling condition is 100 bar pressure and 370°C. The rated Rankine cycle efficiency is 0.3774, and and the parasite losses is about 85%, and thus the net efficiency is 0.82. The steam blow-down fraction is 0.02. NG backup boiler has 0.9 efficiency based on Low Heating Value (LHV). The steam cycle can be operated at maximum to rated ratio of 1.05, or maximum output of 105 MW, and the minimum operation output is 25 MW; and the minimum start up temperature of the steam is 300°C.

The solar field data from the SAM output is given as follows.

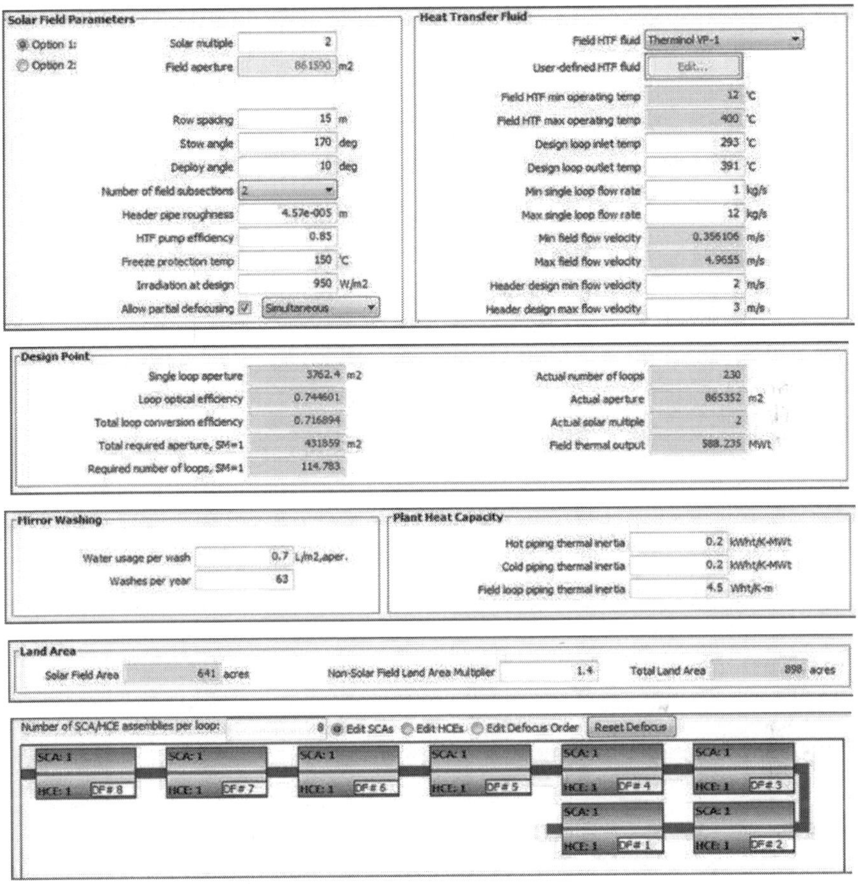

As shown in the result, the Solar Multiple is 2, for 6 hours TES. The term solar multiple (SM) is useful metric to evaluate the performance and solar field sizes. The SM is the ratio of the solar energy collected at the design point to the amount of solar energy required to generate the rated turbine gross power. An SM of 1.0 means that the solar field delivers exactly the amount of energy required to run the plant at its design output, at the design point solar conditions. A larger SM indicates larger solar system. The design point is the reference set of conditions selected for designing the system. For PTC, the design conditions are: cosq = 0, ambient temperature = 25°C, DNI = 1000 W/m^2, and wind speed = 5 m/s. The cosq = 0 means that the incidence angle is zero degrees or the sun is normal to the collector aperture.

Based on the design point conditions in the sections above, solar fields with SM range from 1 to 3.5 have solar field areas as given in the **Table 3**, [15].

Table 3. Solar multiple field areas.

Solar multiple (SM)	1	1.2	1.5	1.8	2.1	2.5	3	3.5
Area m^2/MW	4552	5493	6847	8202	9594	11,399	13,657	15,952

Thus, the aperture area per MW is 8616 m^2/MW. The DNI solar in Cairo is given as 1840 kWh/m^2. This may be lower than that of Doha, Qatar (which is about 2100 kWh/m^2), but the dusty weather in Doha may give DNI less than that of Cairo. The irradiation design is taken as 950 W/m^2.

The solar field thermal output is 588 MW thermal (MWt). This is achieved by using 86,5252 m^{2}total aperture area by using 230 number of loops of 3762.4 m^2 area per loop, and loops total optical efficiency equal to 0.716. The land area for the solar field is 641 acres (2,589,640 m^2), and the land total area is 898 acres (3627, 920 m^2 or 4.63 square kilometer), by assuming that the total land area is 1.4 the solar field area.

The HTF operating condition are: the minimum operating temperature is 12°C, maximum operating temperature is 400°C, design inlet loop is 293°C, and outlet loop minimum temperature is 391°C. The minimum and maximum HTF flow rates in a single loop is 12 kg/ and 1 kg/s respectively. Data on solar SCA are given as:

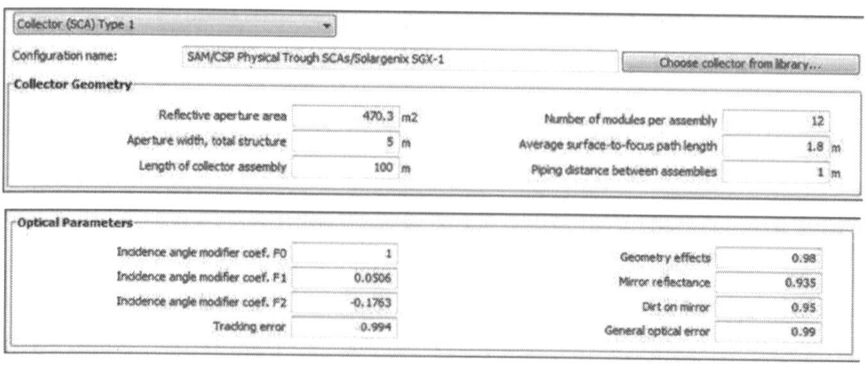

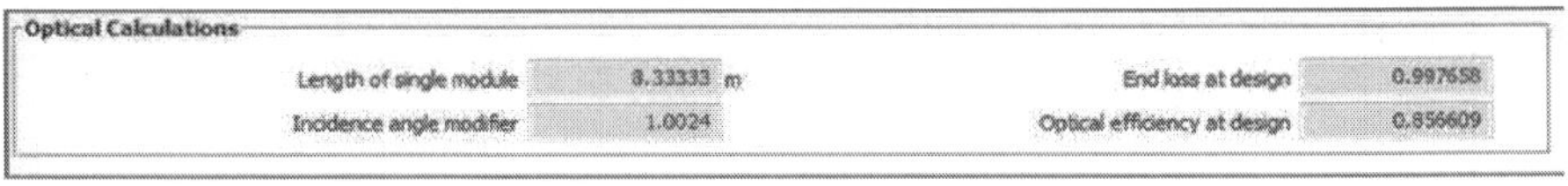

Optical Calculations			
Length of single module	8.33333 m	End loss at design	0.997658
Incidence angle modifier	1.0024	Optical efficiency at design	0.856609

4. CONCERNING THE USED RECEIVERS

The net efficiency of the power block plant (net EP out/ heat rate output of the SG) is 0.32. Thus, heat is to be generated at rate 312.5 MW tin the SG. The ratio of the SG heat output to the solar heat is in the range of 65% - 70%, and thus the required solar heat input is (312.5/0.65=) 481 MWt when no TES is considered.

The storage system data are given as follows:

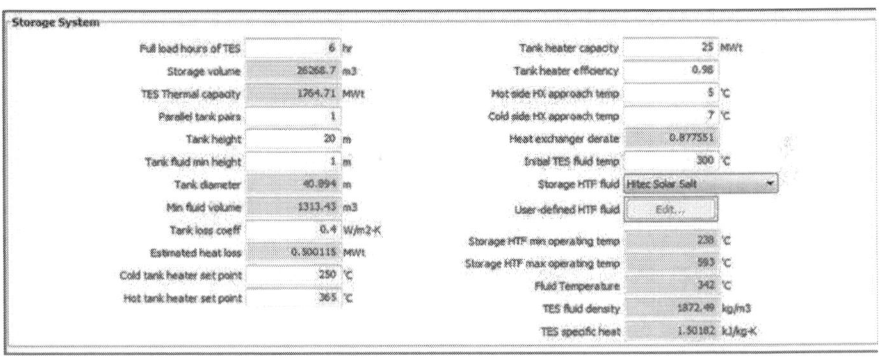

Storage System					
Full load hours of TES	6	hr	Tank heater capacity	25	MWt
Storage volume	26268.7	m3	Tank heater efficiency	0.98	
TES Thermal capacity	1764.71	MWt	Hot side HX approach temp	5	°C
Parallel tank pairs	1		Cold side HX approach temp	7	°C
Tank height	20	m	Heat exchanger derate	0.877551	
Tank fluid min height	1	m	Initial TES fluid temp	300	°C
Tank diameter	40.894	m	Storage HTF fluid	Hitec Solar Salt	
Min fluid volume	1313.43	m3	User-defined HTF fluid	Edit...	
Tank loss coeff	0.4	W/m2-K	Storage HTF min operating temp	238	°C
Estimated heat loss	0.900115	MWt	Storage HTF max operating temp	593	°C
Cold tank heater set point	250	°C	Fluid Temperature	342	°C
Hot tank heater set point	365	°C	TES fluid density	1872.49	kg/m3
			TES specific heat	1.50182	kJ/kg-K

The capital costs include: land side improvement, solar field cost (about \$270/ m^2), HTF cost, TES cost of 1765 MWt (\$80/kWht), the power block (\$830/kW), and the balance of the plant (\$110/kW). The capital costs are given by SAM program as:

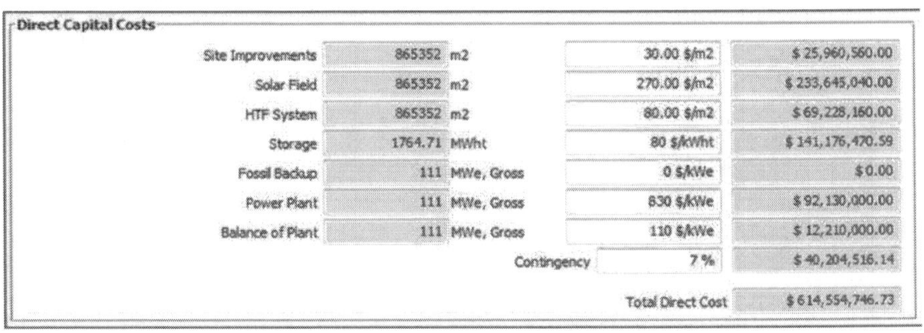

Direct Capital Costs				
Site Improvements	865352	m2	30.00 \$/m2	\$ 25,960,560.00
Solar Field	865352	m2	270.00 \$/m2	\$ 233,645,040.00
HTF System	865352	m2	80.00 \$/m2	\$ 69,228,160.00
Storage	1764.71	MWht	80 \$/kWht	\$ 141,176,470.59
Fossil Backup	111	MWe, Gross	0 \$/kWe	\$0.00
Power Plant	111	MWe, Gross	830 \$/kWe	\$ 92,130,000.00
Balance of Plant	111	MWe, Gross	110 \$/kWe	\$ 12,210,000.00
		Contingency	7 %	\$ 40,204,516.14
			Total Direct Cost	\$ 614,554,746.73

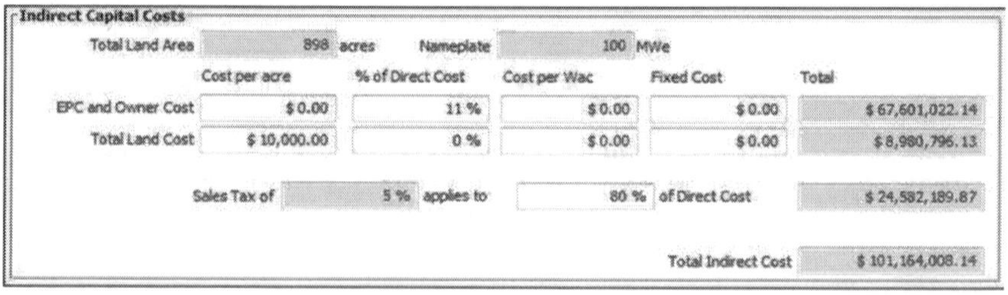

Indirect Capital Costs

Total Land Area 898 acres — Nameplate 100 MWe

	Cost per acre	% of Direct Cost	Cost per Wac	Fixed Cost	Total
EPC and Owner Cost	$ 0.00	11 %	$ 0.00	$ 0.00	$ 67,601,022.14
Total Land Cost	$ 10,000.00	0 %	$ 0.00	$ 0.00	$ 8,980,796.13

Sales Tax of 5 % applies to 80 % of Direct Cost $ 24,582,189.87

Total Indirect Cost $ 101,164,008.14

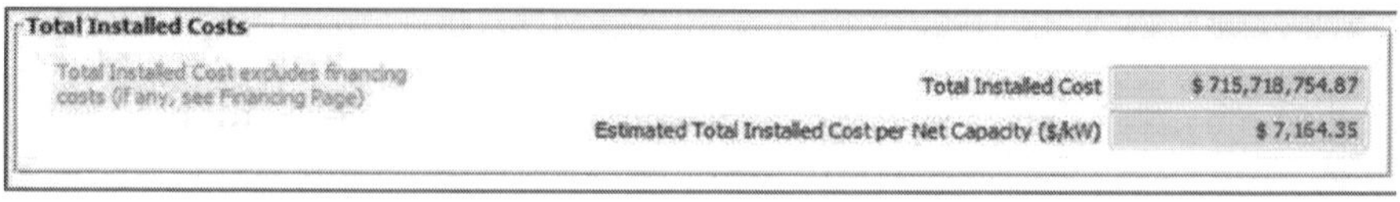

Total Installed Costs

Total Installed Cost excludes financing costs (if any, see Financing Page)

Total Installed Cost $ 715,718,754.87

Estimated Total Installed Cost per Net Capacity ($/kW) $ 7,164.35

The operation costs are given as:

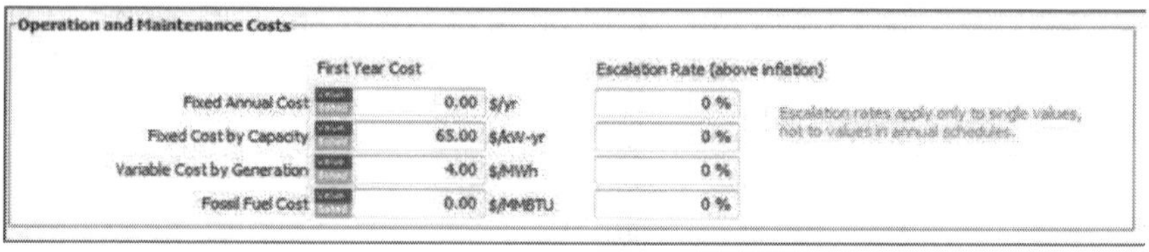

Operation and Maintenance Costs

	First Year Cost		Escalation Rate (above inflation)
Fixed Annual Cost	0.00	$/yr	0 %
Fixed Cost by Capacity	65.00	$/kW-yr	0 %
Variable Cost by Generation	4.00	$/MWh	0 %
Fossil Fuel Cost	0.00	$/MMBTU	0 %

Escalation rates apply only to single values, not to values in annual schedules.

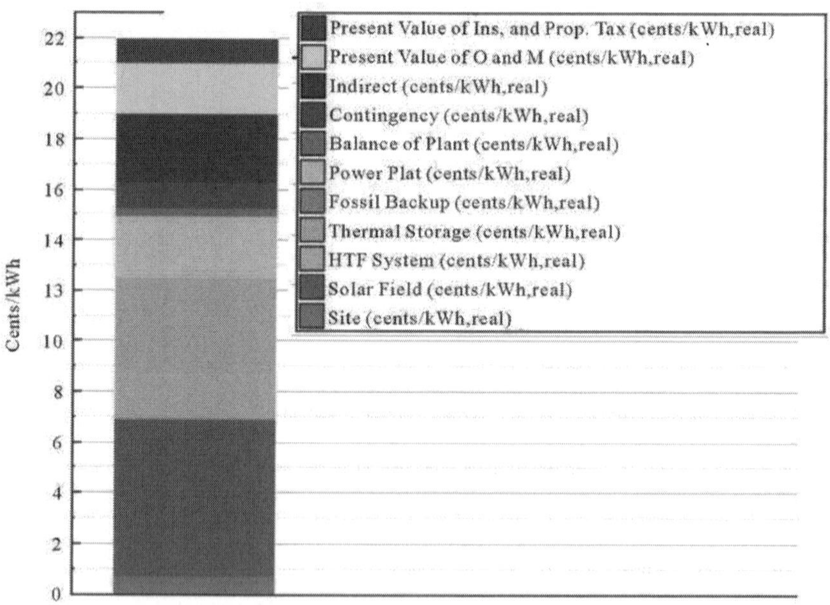

Figure 15. Itemized LEC cost.

The above data were used to calculate the LEC, based on 25 years for the plant life, 2.5% inflation rate, 8.2% real discount rate, and 10.9% nominal discount rate. The itemized LEC is shown in**Figure 15**. The plant electric output is given in **Figure 15**.

The annual energy output is 247.345 GWh, the real LEC is 21.93 cent/kWh, capacity factor is 28.3%, and annual water usage is about 1 million cubic meters (Mm^3/y) by using evaporative cooling condenser. **Figure 16** shows the monthly electric power output of the plant.

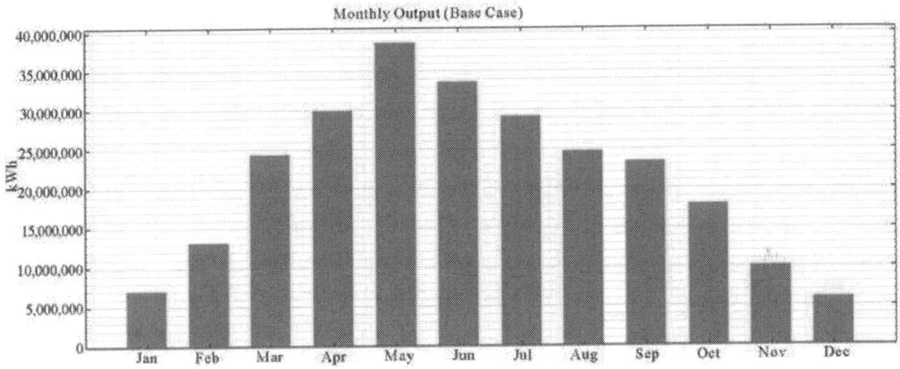

Figure 16. Plant monthly electric power output.

5. CONCLUSIONS

The utility solar power plants were reviewed. The installed capital cost per MW of SPP is much higher than that of conventional plants. Thus, substantial economic support will be required for CSP project economic viability. For the utility scale power production, parabolic trough is only technology that can have operational track record, which gives it a moderate technology risk (low relative to other CSP technologies).

A 100 MW SPP using PTC was considered suitable for Qatar. The result shows that the total capital cost is $7.16 Million (M) per MW, and the LEC is $0.22/kWh.

Linear Fresnel technology can be more competitive with parabolic trough for utility-scale generation, but more projects are needed to build reliability record for large SPP. There is insufficient actual construction experience to confirm costs of utility scale projects at this stage. There are enough experience with LFC for small projects such as solar cooling and desalination. There is insufficient actual construction experience to confirm costs of utility scale projects at this stage.

REFERENCES

1. M. Darwish and R. Mohtar, "Qatar Energy Footprint," Submitted for Publication to the Journal of Energy Strategy.
2. M. Mendelsohn, T. Lowder and B. Canavan, "UtilityScale Concentrating Solar Power and Photovoltaic Projects: A Technology and Market Overview," Prepared under Task No. SM10.2442 Technical Report, NREL/ TP-6A20-51137, 2012.
3. Appendix A, "Key Assumptions for Levelized Cost of Electricity Ranges". http://www.ucsusa.org/assets/documents/clean_energy /Appendix-Key-Assumptions-Levelized-Costs.pdf
4. "Final Report on Technical Data, Costs, and Life Cycle Inventories of Solar Thermal Power Plants," Project No: 502687, NEEDS New Energy Externalities Developments for Sustainability Integrated Project. http://www.needs-project.org/RS1a/RS1a%20D12.2%20 Final%20report%20concentrating%20solar%20thermal%20-power%20plants.pdf
5. Concentrated Thermal Power Now. http://www.greenpeace.org/raw/content/international/press/reports/Concentrated-Solar-Thermal-Power.pdf
6. Utility Scale Solar Power Plants; Guide for Developers and Investors, International Finance Corporation (IFC), World Bank Group, 2012.
7. M. J. Wagner, "Results and Comparison from the SAM Linear Fresnel Technology Performance Model Preprint," 2012 World Renewable Energy Forum, Denver, 13-17 May 2012, p. 2.
8. "Puerto Errado 2 Thermosolar Power Plant (PE2)," Concentrating Solar Projects, National Renewable Energy Laboratory, 2012. http://www.nrel.gov/csp/solarpaces/project_detail.cfm/projectID =159
9. "Planta Solar 20," Concentrating Solar Projects, National Renewable Energy Laboratory, 2012. http://www.nrel.gov/csp/solarpaces/project_detail.cfm/projectID=39

10. NREL Concentrating Solar Project, Maricopa Solar Project (Maricopa) Plant. http://www.nrel.gov/csp/solarpaces/project_detail.cfm/projectID=58
11. P. R. Fraser, "Stirling Dish System Performance Prediction Model," M.Sc. Thesis, Mechanical Engineering Department, University of Wisconsin, Madison, 2008.
12. NREL Concentrating Solar Project, Maricopa Solar Project (Maricopa) Plant. http://www.nrel.gov/csp/solarpaces/project_detail.cfm/projectID=58
13. 24 MW, SinAn, South Korea. http://www.conergy.com.au/PortalData/1/Resources/master/images/references/freefield/Solar_farm_SinAn.pdf
14. AM0100, "Integrated Solar Combined Cycle (ISCC) Projects—Version 1.0.0". https://cdm.unfccc.int/methodologies/DB/BES7OQGMZYOMCP9JPTVJHP93BVK4UO
15. M. Darwish, H. K. Abdulrahim and R. Mohtar, "Solar Power Desalting Plant in Qatar: Design and Analysis," LAP LAMBERT Academic Publishing, Saarbrücken, 2012.
16. P. Gilman and A. Dobos, "System Advisor Model, SAM 2011.12.2: General Description," Technical Report, NREL/ TP-6A20-53437, 2012. http://www.nrel.gov/docs/fy12osti/53437.pdf

4

CHAPTER

SOLAR ENERGY SUSTAINABILITY IN JORDAN

Ahmad Qasaimeh[1], Mohammad Qasaimeh[2], Zaydoun Abu-Salem[3], Mohammad Momani[4]

[1]*Department of Civil Engineering, Jerash University, Jerash, Jordan*

[2]*Chemical Engineering Department, AlHuson University College, Al-Balqa Applied University, Salt, Jordan*

[3]*Department of Civil Engineering, Philadelphia University, Amman, Jordan*

[4]*Department of Electrical Engineering, Yarmouk University, Irbid, Jordan*

ABSTRACT

Jordan is a country with highly fossil fuel deficiency and thus other energy sources are needed to be explored. Solar energy in Jordan is highly recognized as a good source of energy and an excellent substitute to the fossil fuel. The solar energy in this article is obtained via data bases and modeling techniques for the specified place coordinate and angle of inclination. The angles of sun irradiations are different throughout the year; therefore solar energy needs to be magnified by optimizing the angle of inclination of solar cells. In this research, the optimized angles throughout the year are obtained to be in the range: 10° - 60°. Solar energy can serve the residential building, the findings of this research show that every 1 m^2 of the solar cell may contribute to about 60% - 70% of customer needs of electricity throughout the year. The application of solar energy concept in the design of building will play an important role in energy sustainability.

Keywords: Solar Energy, Inclination Angle, Sustainable Buildings

1. INTRODUCTION

During the last two decades, the increasing energy demand has brought challenges to Jordan due to country's limited resources. Hence, solar energy applications have more attention to substitute the depletion of the fossil fuel that causes dramatic pollution to the environment. Jordan has established a strategic change and reform of its national economy and energy strategy [1] . Jordan has assisted programs utilizing solar energy. Assessment involved systematic monitoring of implementation of appropriate technologies, demonstrations, and pilot projects [2] [3] .

Due to high and reliable solar irradiance in Jordan (5.5 $kWh/m^2 \cdot d$), the usage for solar energy in Jordan has high potential for about 330 sunny days per year using solar panels [4] . Solar radiation also differs according to seasons, in winter the sun becomes lower in the sky and higher in summer because sun ray's angle changes due to the earth's tilt angle [5] .

The distribution of total radiation on a horizontal surface over a day was examined by Liu and Jordan who showed that the ratio of hourly to daily radiation could be correlated with the local day length and angle which differs through the year [6] . Solar energy encounters many parameters that affect its cultivation during the year such as sunshine duration, relative humidity, temperature, and cloudiness. The solar panels inclination therefore should be dynamically changed during the seasons at certain place [7] .

However solar radiation differs along the seasons. The results of Liu and Jordan were confirmed by Collares Pereira and Rabl [6] using a wider database for the average distribution of solar radiation associated with different coordinates of time and place. Saraf and Hamad [8] found the yearly optimum tilt angle in Basra, Iraq was higher than the latitude by about 8°. Gopinathan [9] showed the optimum tilt angle of oriented sloping plates is almost equal to place latitude.

Both Gopinathan [10] and Soulayman [11] showed the optimum tilt angle of oriented sloping plates is almost equal to place latitude.

However, researchers have different approaches for optimal angles for solar collectors in different places, because the radiation pattern changes from location to location and time to time [12] [13] . Thus, the aim of this research is to optimize the angle of inclination for Jordan during the year.

2. METHODOLOGY

The angles of sun irradiations are different throughout the year. Therefore, the solar panels should be dynamically inclined with different angles. This article spots the light on solar energy utilization depending on solar energy databases and modeling techniques. The information about solar energy, temperature, and electricity consumption was collected from several organizations namely: National Center for Research and Development, Ministry of Energy and Mineral Resources, and Jordan Meteorological Department. The sunshine hours in Jordan zone were taken from the time and date calendar. Solar Energy Modeling was performed using Meteonorm software Version 6 for modeling solar energy with inclination angle of panels for Jordan database.

3. SOLAR ENERGY FOR BUILDINGS IN JORDAN

The major goal of this research is to explore that the collected solar energy can offset the electrical energy consumption in residential buildings (Table 1) [14] . The aim of the research can be achieved via many tracts. The first tract is to optimize solar panel angle of inclination throughout the year. The second tract is to design sufficient area in the residential buildings for panels to be installed. The third tract is to manage the energy in the building in the basis of building design and geometry, and daily wise management of energy.

Fossil fuel depletes and costly increases with the time, furthermore it causes environmental problems such as global warming. Therefore, solar cells must be oriented and distributed effectively depending on time in the year and depending on the building size. The savings of electricity can be enhanced by altering the daily time of wakeup and sleep. The daylight hours may help in utilizing the natural sunlight instead of electricity [15] . The implementation of Daylight Saving Time (DST) creates an additional hour of higher outdoor air temperature and solar radiation during the primary cooling times of the evening [16] . California Energy Commission [17] [18] conducted a simulation-based study to examine the effects of DST on statewide electricity consumption. Consequently, by concise management, collecting sun irradiation and fitting the daily man activities to sunshine will compensate large part of electricity for residential building.

The records about the solar energy in Jordan spots the light on the truth that Jordan is rich in sun irradiations as Table 2 shows the data about sunshine hours and solar energy in different places in Jordan in the year [19] . The average of the data values shows the relative trend between solar energy and sunshine period as it is shown in Figure 1. The solar energy data accumulates an estimated average energy value on the yearly basis of about 2056 kWh/m^2 for Jordan.

In Figure 1, the information obtained about the solar energy can be represented via analytical model that incorporates the sunshine period as a major parameter. After calibration, this model can forecast the energy for specific place coordination and variable angle of inclination.

Table 1. The annual electricity consumption per capita in residential building in Jordan

Equipment	Service (hours/day)		Operation	Consumption	Percent of Consumption
	Summer	Winter	(hours/year)	(kWh)	
Lighting	7	6	2340	514.8	21.28%
Refrigerator	16	8	4392	933.3	38.58%
TV	12	7	3372	288.3	11.92%
W. Machine	2	2	220	89.1	3.68%
Iron	1	1	130	117	4.83%
Fan	14	0	420	23.5	0.97%
Water Pump	1.5	1.5	135	10.8	0.45%
Freezer	10	2	2160	113.4	4.69%
Water Cooler	2	0	90	0.9	0.04%
Vacuum Cleaner	1	1	120	24	0.99%
Washing Dryer	0.5	0.5	180	5.4	0.22%
Hear Dryer	0.5	0.5	50	7.5	0.31%
Heater	1	1	60	2.7	0.11%
Geyser	1	5	480	192	7.94%
Air Condition	10	0	350	96.3	3.98%
			Total	**2419**	**100%**

4. SOLAR ENERGY MODELING THROUGHOUT THE YEAR

In this section, the solar energy in Jordan is being modeled for different angle of solar panel inclination. Meteonorm version 6 uses data bases about the specified place coordination and gives the estimated solar energy for different panel inclinations. It comprises physical and environmental parameters applicable for certain coordinate. The estimation of solar energy is being adjusted for different angle of orientation of solar panel due to sun movement during the months of the year. The modeling process is shown on Table 3 that depicts the solar energy throughout the months in the year for different solar cell inclination.

Table 2 . Data of sun shine period (hours) and energy (kWh/m^2) during the report in 2007 in different stations in Jordan.

Sunshine	Amman	Irbid	Dhlail	Azraq
JAN	5.1	5.5	5.5	6
FEB	5.8	5.2	6.5	6.6
MAR	7.7	7.5	8.6	8.4
APR	9.4	9	9.7	9.3
MAY	9.8	9.6	9.9	9.6
JUN	11.8	11.3	11.7	11.1
JUL	12	11.8	12	11.2
AUG	11.7	12.2	11.6	11.1
SEPT	10	10	11	9.7
OCT	8.8	8.5	9	8.7
NOV	6.2	6.3	6.7	7
DEC	5	4.2	5.7	5.5

Energy	Amman	Irbid	Dhlail	Azraq
JAN	3.2	3.1	3.2	3.3
FEB	4.1	3.7	4.1	4.2
MAR	5.4	5.1	5.5	5.7
APR	6.8	6.4	6.8	6.8
MAY	7.1	6.9	7.5	7.5
JUN	8.3	7.7	8.4	8.4
JUL	8.1	7.6	8.3	8.1
AUG	7.4	7.5	7.7	7.3
SEPT	6	6	6.5	6.2
OCT	4.5	4.6	4.9	4.9
NOV	3.3	3.3	3.3	3.7
DEC	2.9	2.9	2.9	2.9

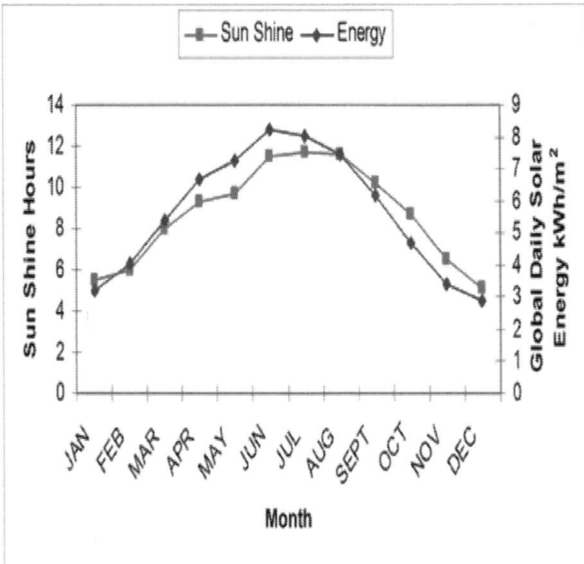

Figure 1. Average sunshine hours and global daily solar energy recorded in Jordan.

Table 3. The estimated solar energy kWh/m² during months of the year for different solar cell inclination.

	Solar Energy kWh/m²					
	10°	20°	30°	40°	50°	60°
JAN	119	136	149	158	163	164
FEB	123	135	143	148	149	147
MAR	175	184	189	189	185	176
APR	204	207	205	198	186	170
MAY	241	236	225	209	189	163
JUN	254	244	228	207	182	153
JUL	262	253	239	219	194	164
AUG	245	245	238	226	208	185
SEP	216	226	229	227	218	204
OCT	185	203	215	221	222	216
NOV	136	154	169	178	183	183
DEC	111	127	140	149	155	156

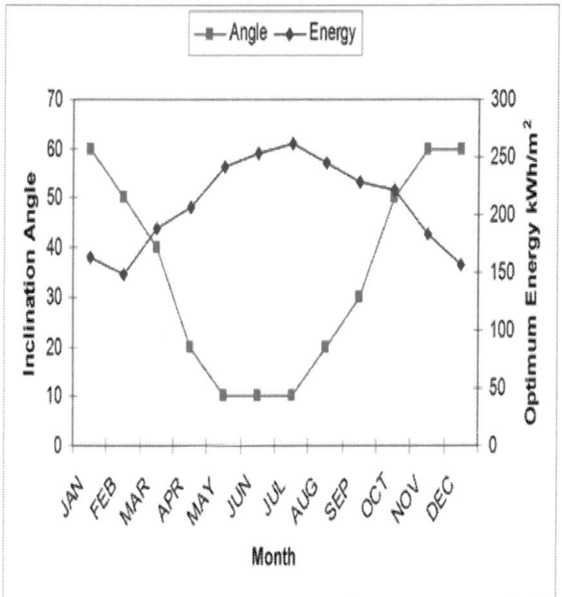

Figure 2. Optimum solar energy with optimum solar cell inclination in each month in Jordan.

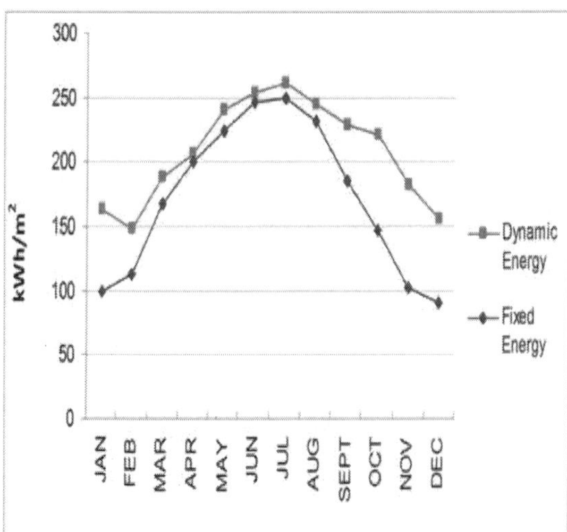

Figure 3. The comparison between energy gathered by panels of dynamic angle of inclination and panels of fixed angle of inclination.

Table 4. The optimal solar energy cultivated during months in Jordan.

MONTH	Inclination Angle	Optimal Solar Energy (kWh/m^2)
JAN	60	164
FEB	50	149
MAR	40	189
APR	20	207
MAY	10	241
JUN	10	254
JUL	10	262
AUG	20	245
SEP	30	229
OCT	50	222
NOV	60	183
DEC	60	156
	Total Energy	2501

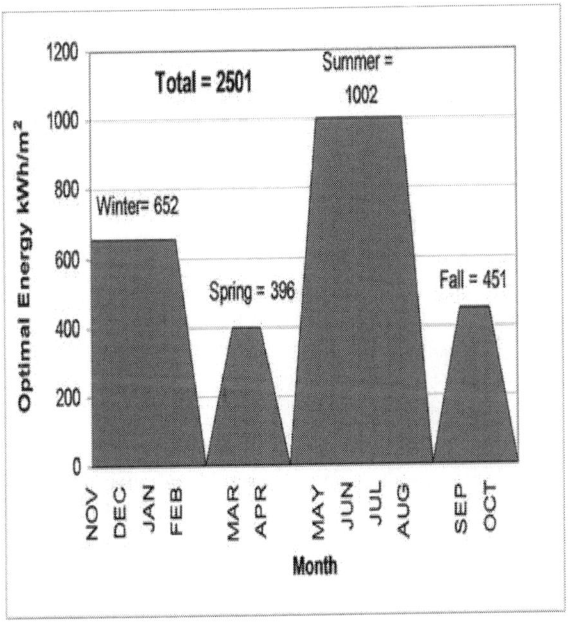

Figure 4. The optimal solar energy cultivated in each season in Jordan.

The energy values in Table 3 can be optimized to maximum solar energy for each month, for example for January the maximum energy is obtained when the panel is inclined to 60°. Figure 2shows the optimum solar energy with optimum solar cell inclination in each month in Jordan.

In this research, the focus is to optimize solar energy cultivation along the year. Hence, the angle of inclination is given the great attention. The data recorded about solar energy is given via panels of fixed inclination (Table 2). Figure 3 compares between the energy recorded data for fixed panels and the dynamic estimated energy values for variable angle of inclination during the year.

The variable angle of inclination represents other important parameter of modeling in addition to the sunshine period. As the sun moves in orbital track, the sunshine period and the sun irradiations direction are variable along the year, which creates the seasons. In Jordan, the year is classified into four seasons: Summer, fall, winter, and spring. The summer season extends from May until August. The fall is characterized to be in September and October. Winter extends from November until February, and spring denotes to March and April. Figure 4 represents each season and the solar energy accumulated with it within variable angle of inclination. For example, the figure shows that the solar energy gained in summer is as high as 1002 kWh/m^2.

As per for Figure 4, the optimized energy upon yearly basis is computed to be 2501 kWh/m^2 and this is also illustrated in Table 4. This value overcomes the value of general solar fixed cells (2056 kWh/m^2) that seen in Figure 1. By comparing the total energy value (2501 kWh/m^2—Table 4) and the value of electricity consumption for residential building per year for single customer (2419 kWh/capita—Table 1); it is worthy to take in consideration that if the solar energy is efficiently converted to electricity then every 1 m^2 of the solar cell will contribute to about 60% - 70% of customer needs of energy in the year. Consequently, the area occupied by solar panels is another important parameter in energy estimation. Hence, the idea of utilizing solar energy in buildings is an important scenario for sustainability.

5. CONCLUSIONS

Jordan is a place that it is recognized of plentiful high solar radiation. The solar energy depends on the sun travel along the years, and thus the sunshine period and sun irradiations direction are variable during the year. Solar energy cultivation is optimized considering dynamic variation of the angle of solar cell inclination through the year.

In this research, it's shown that solar energy can serve the residential building consumption of electricity as every 1 m^2 of the solar cell may contribute to about 60% - 70% of customer needs of energy in the year. In addition, combining solar energy concept in the design of building will lower the cost of energy use and will play an important role in sustainable development of buildings.

REFERENCES

1. Abdelkader, M.R., Al-Salaymeh, A., Al-Hamamre, Z. and Sharaf, F. (2010) A Comparative Analysis of the Performance of Monocrystalline and Multiycrystalline PV Cells in Semi Arid Climate Conditions: The Case of Jordan. Jordan Journal of Mechanical and Industrial Engineering, 4, 543-552.
2. Blumenberg, J., Bentenrieder, M. and Kerschensteiner, H. (1997) Introducing Advanced Testing Methods for Domestic Hot Water Storage Tanks in Jordan. Renewable Energy, 10, 207-211.
3. Badran, A. (2001) A Study in Industrial Applications of Solar Energy and the Range of Its Utilization in Jordan. Renewable Energy, 24, 485-490.
4. Shariah, A., Al-Akhras, M.-A. and Al-Omari, I.A. (2002) Optimizing the Tilt Angle of Solar Collectors. Renewable Energy, 26, 587-598.
5. Jibril, Z. (1991) Estimation of Solar Radiation over Jordan Predicted Tables. Renewable Energy, 1, 277-291.
6. Collares-Pereira, M. and Rabl, A. (1979) The Average Distribution of Solar Radiation-Correlations between Diffuse and Hemispherical and between Daily and Hourly Insolation Values. Solar Energy, 22, 155-164.
7. Singh, H.N. and Tiwari, G.N. (2005) Evaluation of Cloudiness, Haziness Factor for Composite Climate. Energy, 30, 1589-1601.
8. Saraf, G.R. and Hamad, F.A.W. (1988) Optimum Tilt Angle for a Flat Plate Solar Collector. Energy Conversion and Management, 28, 185-191.
9. Gopinathan, K.K. (1988) A General Formula for Computing the Coefficients of the Correlation Connecting Global Solar Radiation to Sunshine Duration. Solar Energy, 41, 499-502.
10. Gopinathan, K.K. (1991) Solar Radiation on Variously Oriented Sloping Surfaces. Solar Energy, 47, 173-179.

11. Soulayman, S.Sh. (1991) On the Optimum Tilt of Solar Absorber Plates. Renewable Energy, 1, 551-554.
12. Gunerhan, H. and Hepbasli, A. (2007) Determination of the Optimum Tilt Angle of Solar Collectors for Building Applications. Building and Environment, 42, 779-783.
13. Yakup, M.Ab.H.M. and Malik, A.Q. (2001) Optimum Tilt Angle and Orientation for Solar Collector in Brunei Darussalam. Renewable Energy, 24, 223-234.
14. Ministry of Energy and Mineral Resources, MEMR (1996) Analytical Study Report. Jordan.
15. Momani, M.A., Yatim, B. and Mohd Ali, M.A. (2009) The Impact of the Daylight Saving Time on Electricity Consumption—A Case Study from Jordan. Energy Policy, 37, 2042-2051.
16. Shimoda, Y., Asahi, T., Taniguchi, A. and Mizuno, M. (2007) Evaluation of City-Scale Impact of Residential Energy Conservation Measures Using the Detailed End-Use Simulation Model. Energy, 32, 1617-1633.
17. Kandel, A. and Metz, D. (2001) The Effects of Daylight Saving Time on California Electricity Use. California Energy Commission (CEC), Sacramento.
18. Kandel, A. (2001) Electricity Savings from Early Daylight Saving Time. California Energy Commission (CEC), Sacramento.
19. Jordan Meteorological Department (JMD) (2007) Jordan Annual Climate Bulletin. Jordan Meteorological Department, Amman.

5

CHAPTER

FUEL PRODUCTION USING CONCENTRATED SOLAR ENERGY

Onur Taylan[1] and Halil Berberoglu[1]

[1] Department of Mechanical Engineering, The University of Texas at Austin, Austin, TX, USA

1. INTRODUCTION

Limited reserves of fossil fuels and their negative environmental effects impose significant problems in our energy security and sustainability. Consequently, researchers are looking for renewable energy sources, for instance solar energy, to meet the energy demands of a growing world population. However, terrestrial solar energy is a dilute resource per footprint area and is intermittent showing substantial variability depending on the season, time of the day, and location.

One strategy to overcome these drawbacks of solar energy is to concentrate and use it for cleaning and upgrading dirty fuels such as coal and other hydrocarbons or converting renewable feedstocks such as biomass into carbon-neutral solar fuels. In this way, the intermittent and dilute solar energy can be concentrated and stored as a chemical fuel which can be easily integrated to our existing energy infrastructure. These advantages of solar fuels produced with concentrated solar radiation make them an attractive solution in our quest for renewable and clean fuels. Figure 1 shows the energy potential and carbon emissions by most commonly used fuels along with solar hydrogen.

Most common and available methods for solar fuel production are thermolysis, cracking, reforming, gasification and through thermochemical cycles. All these methods require high temperatures to produce solar fuel. Therefore, in these methods, there are some qualities of the feedstock or the reactor that need to be satisfied to attain high temperatures and efficient solar fuel production. For instance, the physical size and porosity of the feedstock play an important role. As the surface area-to-volume ratio of the feedstock increases, more reaction sites will be available for the reaction to occur, which increases the process efficiency. The feedstock should also have a narrow bad gap to lower the energy requirement for chemical process. Additionally, the material on the reactor walls should have high optical absorption to increase the temperature of the reactor and withstand high temperatures, and the window material should have high transmissivity to let the solar energy in to the reactor. More detailed property requirements are given by Nowotny *et al.* [1].

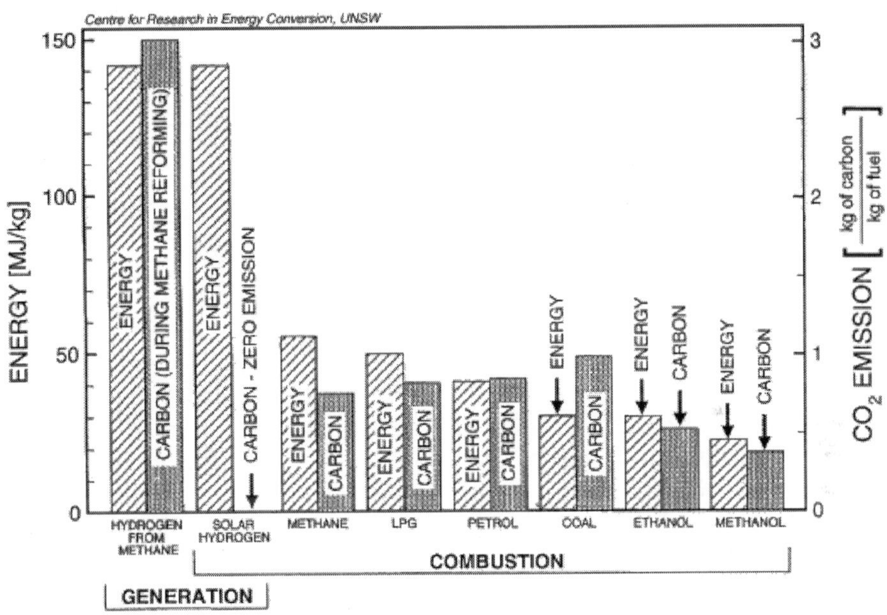

FIGURE 1. Comparison of different fuels in terms of their energy produced and CO_2 emission [1].

This review chapter consists of four sections. Following the introduction, the second section "Concentrated Solar Fuel Production Methods" reviews the different routes of producing solar fuels according to the feedstock material used in the processes. These include (i) thermolysis of water, (ii) thermochemical cycles, (iii) cracking of gaseous hydrocarbons, and (iv) gasification and reforming of coal and biomass. These methods are compared with each other based on their temperature, pressure, thermodynamic efficiencies, and by-products. The third section "Concentrated Solar Reactors" provides a comprehensive review of different concentrated solar reactor designs reported in the literature. This section first reviews the current solar concentration methods and describes in detail the effects of concentrating factors on the heat flux and temperatures that can be achieved. Then, the section describes the design and basic principles of operation of different solar reactors, their applicability for the different methods described in the preceding section, and their temperature and pressure capabilities. Moreover, the section summarizes the reported solar to fuel conversion efficiencies of each design. Finally, the chapter ends with the conclusions and outlook of fuel production with concentrated solar energy outlining the challenges, new research directions and novel applications.

2. CONCENTRATED SOLAR FUEL PRODUCTION METHODS

This section describes different methods of producing solar fuels according to the feedstock material used in the respective processes.

2.1. Thermolysis of Water

The term "thermolysis of water" refers to the thermal decomposition of water molecules into hydrogen and oxygen gases. Historically, due to high availability and simple molecular form of water, researches on solar fuel production started with direct hydrogen production by thermolysis of water using solar energy as,

$$H_2O \rightarrow H_2 + \frac{1}{2} O_2 \qquad \Delta H_{300K} = 286 \ \frac{kJ}{mol} \tag{1}$$

The reaction given in Equation (1) is an endothermic process, i.e., it requires energy to break the bonds. However, breaking all the bonds in water molecules requires temperatures as high as 2500 K [2]. At lower temperatures, partial decomposition occurs. Although it is possible to reach

2500 K with concentrated solar energy, the reactor where this process takes place shows material issues related to high temperatures. Additionally, after the dissociation of water molecules, hydrogen and oxygen gases require separation at high temperatures in order to prevent back-bonding, i.e., reproduction of water molecules with an exothermic process. Some solutions include cooling the reactor down by injecting a gas or expanding these gases through nozzle at the end of the reactor [2, 3]. Other solutions include using double or tubular membranes or using multi-stage steam ejectors to lower the exit pressure [4]. However, these solutions further reduce the efficiency of the process, and thus no commercial plant using this technology exists.

2.2. Thermochemical Cycles

Some metal oxides are reduced using solar energy since metals provide good storage and transport of energy, such as solar energy. Such metal oxides include, but not limited to ZnO, MgO, SnO_2, CaO, Al_2O_3 and Ce_2O_3. The reduction step of these metal oxides is generally followed by an oxidation step at lower temperatures than reduction step in order to produces solar fuel, mainly hydrogen. The reduced metal oxides generally react with CO_2 or steam. If steam is used in oxidation that step is called hydrolysis. The thermochemical cycles of different metal oxides are generally compared based on their temperature requirements for the reduction step, the reaction or dissociation rates and reaction kinetics.

ZnO is one of the most popular oxides mainly due to its abundance and relatively low temperature requirement for complete dissociation when compared to other metal oxides. Additionally, since ZnO is a simple metal oxide, it does not undergo multiple reactions before its full dissociation. The dissociation of ZnO occurs as according to,

$$\begin{gathered} ZnO \rightarrow Zn + {}^{1}\!/_{2}O_2 \\ \Delta H_{2000K} = 546\ {}^{kJ}\!/_{mol} \end{gathered} \tag{2}$$

The complete dissociation of ZnO to Zn requires temperatures higher than about 2300 K whereas, for instance, the dissociation of MgO as another simple metal oxide requires about 3700 K at atmospheric pressures [3, 5]. As in water thermolysis, partial dissociations can occur at lower temperatures. Although hydrolysis of zinc is exothermic as given by Equation (3), only 24% of Zn could be oxidized to produce H_2 at a reactor temperature of 800 K and an atmospheric pressure [6].

$$Zn+H_2O \rightarrow ZnO+H_2 \qquad \Delta H_{300K}=-62\ {}^{kJ}\!/_{mol} \tag{3}$$

Figure 2 shows the overall process of hydrogen production from zinc-oxide.

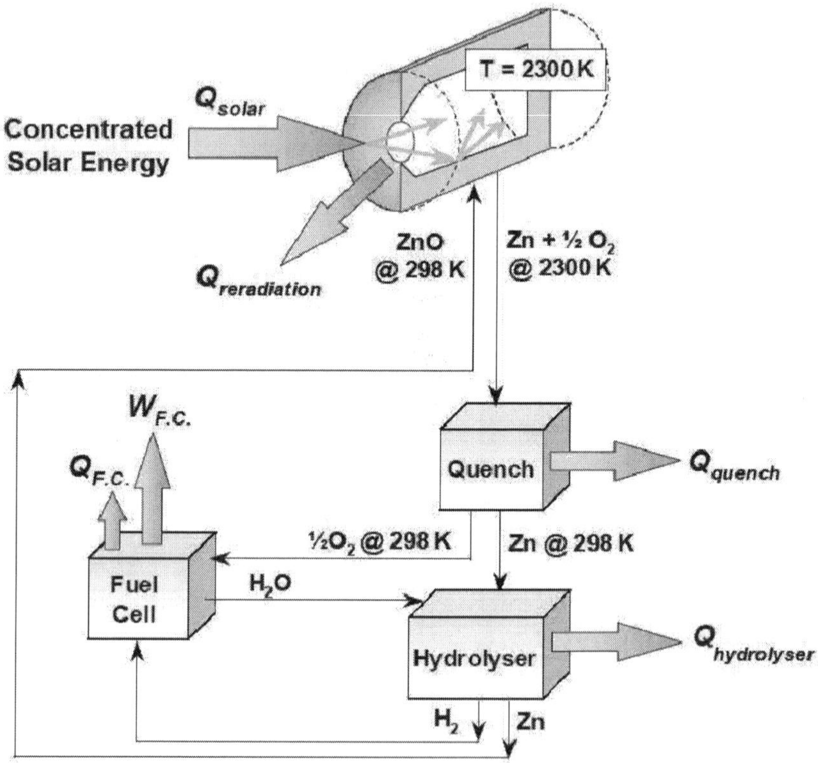

FIGURE 2. Flowchart for thermochemical hydrogen production from zinc-oxide using concentrated solar energy [5].

As an alternative to ZnO reduction, Abanades *et al.* [7] proposed SnO_2 reduction using solar energy. Once the SnO_2 is reduced to SnO in gaseous form using solar energy at temperatures nearly 1600°C, hydrolysis of SnO with steam at about 550°C and ambient pressure takes place in another step to form hydrogen gas as,

$$SnO_{2(s)} \rightarrow SnO_{(g)} + \frac{1}{2} O_2 \qquad \Delta H_{1873K} = 557 \text{ kJ/mol} \tag{4}$$

$$SnO_{(s)} + H_2O_{(g)} \rightarrow SnO_{2(s)} + H_2 \qquad \Delta H_{773K} = -49 \text{ kJ/mol} \tag{5}$$

The advantages of SnO_2/SnO reduction when compared to ZnO/Zn reduction are that (i) the SnO_2-to-SnO conversion can be increased in Equation (4) by decreasing the pressure of the solar reactor which increases the overall conversion efficiency [7] (ii) SnO has higher melting and boiling points when compared to those of Zn, so that quenching rate of SnO is not as important as of Zn [7] (iii) in ZnO/Zn dissociation, Zn needs to be quenched rapidly below its condensation temperature to prevent recombination, while this is not the case with SnO_2/SnO system.

There are some other metals that can be reduced with faster reaction kinetics such as Ce_2O_3. However, the reduction of Ce_2O_3 to CeO_2 starts at temperatures higher than 2300 K [8, 9]. Full dissociation requires higher temperatures. This requirement of high temperatures creates some material limitations on the material of the reactor and increases the cost of the reactor significantly. Although there are some lab-scale prototypes of Ce_2O_3/CeO_2 solar reactor, it is not preferred due to these limitations and high cost.

Another research was also started with producing hydrogen gas from hydrogen sulfide, H_2S, as,

$$H_2S \rightarrow H_2 + \frac{1}{2} S_2 \qquad \Delta H_{300K} = 91.6 \text{ kJ/mol} \tag{6}$$

Hydrogen sulfide is a toxic by-product gas of sulfur removing process from natural gas, petroleum and coal. Thermal decomposition of hydrogen sulfide requires about 1800 K [10]. It is advantageous over the other metal oxide reduction processes discussed above since this thermochemical process is only a one-step process that does not require additional oxidation step to produce hydrogen. Additionally, the temperature requirement for dissociation is lower than that for the direct water thermolysis. However, the product gases need to be cooled down after the dissociation as in the water thermolysis or other metal oxide reduction processes [11]. Some studies showed that the temperature of reduction could be reduced to about 1500 K, and they showed that the reproduction of hydrogen sulfide is unimportant below 1500 K [3, 12, 13].

In general, the solar chemical process is a clean way to produce hydrogen without any carbon prints. Therefore, the hydrogen as a product of the solar chemical process can be used in fuel cells directly as it is pure. The solar chemical reduction step of the process produce nanoparticles with high surface area to volume ratio, e.g., Zn, SnO which also create additional reaction centers for the hydrolysis to occur [7]. Therefore, the oxidation or hydrolysis occurs fast due to high mass transport of gases in the solid phase [7]. As in the other dissociation processes, the products of the dissociation also need to be cooled in order to prevent re-oxidation. Sandia National Laboratories of US released a comprehensive report on the thermochemical cycle selection with initial selection for further research [14], and Table 1summarizes the studied thermochemical cycles [15].

2.3. Cracking of Gaseous Hydrocarbons

The term “solar thermal cracking” or “solar cracking” is used for thermal decarbonization of natural gas or other hydrocarbons. As a result of cracking, hydrogen, carbon and other possible products are formed without CO_2 emissions. Therefore, this process is another method for clean fuel production. Cracking requires high temperatures of about 1500 K [16] that can be reached using concentrating solar collectors. For example, Maag *et al.* [17] tested a concentrated solar collector with a concentrating factor of 1720, and obtained a maximum temperature of 1600 K within the solar cavity reactor. In general, the advantages of solar cracking are the increase in value of feedstock using solar energy, pure and uncontaminated products and no CO_2 emission [16].

As being the simplest hydrocarbon and the main constituent of natural gas as given in Table 2, methane has been mainly considered for solar cracking. Chemical reaction of evolution of carbon black and methane is given in equation (7) [18, 19]. The kinetic mechanism of methane cracking at 1500 K and atmospheric pressure was proposed as [20, 21],

TABLE 1. Summary of Thermochemical Cycles [15].

Cycle	Reaction Steps
High Temperature Cycles	
Zn/ZnO	$Fe_3O_4 \xrightarrow{2000-2300^\circ C} 3FeO + \frac{1}{2}O_2$ $3FeO + H_2O \xrightarrow{400^\circ C} Fe_3O_4 + H_2$
FeO/Fe_3O_4	$CdO \xrightarrow{1450-1500^\circ C} Cd + \frac{1}{2}O_2$ $Cd + H_2O + CO_2 \xrightarrow{350^\circ C} CdCO_3 + H_2$ $CdCO_3 \xrightarrow{500^\circ C} CO_2 + CdO$
Cadmium carbonate	$CdO \xrightarrow{1450-1500^\circ C} Cd + \frac{1}{2}O_2$ $Cd + 2H_2O \xrightarrow{25^\circ C, electrochemical} Cd(OH)_2 + H_2$ $Cd(OH)_2 \xrightarrow{375^\circ C} CdO + H_2O$
Hybrid cadmium	$Mn_2O_3 \xrightarrow{1400-1600^\circ C} 2MnO + \frac{1}{2}O_2$ $2MnO + 2NaOH \xrightarrow{627^\circ C} 2NaMnO_2 + H_2$ $2NaMnO_2 + H_2O \xrightarrow{25^\circ C} Mn_2O_3 + 2NaOH$
Sodium manganese	$Fe_{3-x}M_xO_4 \xrightarrow{1200-1400^\circ C} Fe_{3-x}M_xO_{4-y} + \frac{y}{2}O_2$ $Fe_{3-x}M_xO_{4-y} + yH_2O \xrightarrow{1000-1200^\circ C} Fe_{3-x}M_xO_4 + yH_2$
M-Ferrite(M = Co, Ni, Zn)	$H_2SO_4 \xrightarrow{850^\circ C} SO_2 + H_2O + \frac{1}{2}O_2$ $I_2 + SO_2 + 2H_2O \xrightarrow{100^\circ C} 2HI + H_2SO_4$ $2HI \xrightarrow{300^\circ C} I_2 + H_2$

Low Temperature Cycles	
Sulfur-Iodine	$H_2SO_4 \xrightarrow{850^\circ C} SO_2 + H_2O + \frac{1}{2}O_2$ $SO_2 + 2H_2O \xrightarrow{77^\circ C\ electrochemical} H_2SO_4 + H_2$
Hybrid sulfur	$Cu_2OCl_2 \xrightarrow{550^\circ C} 2CuCl + \frac{1}{2}O_2$ $2Cu + 2HCl \xrightarrow{425^\circ C} H_2 + 2CuCl$ $4CuCl \xrightarrow{25^\circ C\ electrochemical} 2Cu + 2CuCl_2$ $2CuCl_2 + H_2O \xrightarrow{325^\circ C} Cu_2OCl_2 + 2HCl$
Hybrid copper chloride	$2CH_4 \rightarrow C_2H_6 + H_2 \rightarrow C_2H_4 + 2H_2 \rightarrow C_2H_2 + 3H_2 \rightarrow 2C_{(solid)} + 4H_2$

TABLE 2. Compositions of Natural Gas from Different Sources [20].

	Volume Fractions (%)					
	CH_4	C_2H_6	C_3H_8	C_4H_{10}	CO_2	N_2
Methane	100	-	-	-	-	-
Modified Algeria Gas	91.2	6.5	2.1	0.2	-	-
Modified Groningen Gas	83.5	4.7	0.7	0.2	-	10.8
North Sea Gas	88.2	5.4	1.2	0.4	1.4	3.2

$$2CH_4 \rightarrow C_2H_6 + H_2 \rightarrow C_2H_4 + 2H_2 \rightarrow C_2H_2 + 3H_2 \rightarrow 2C_{(solid)} + 4H_2 \quad (7)$$

Another important aspect of producing hydrogen and carbon black (solid carbon) is their market values. Hydrogen and carbon black have a market value of about $135 billion per year and between $7 and $11 billion per year depending on the grade of the carbon black in the world, respectively [22].

Under an EU project named SOLHYCARB, a 50-kW$_{th}$ indirectly heated, cavity type solar reactor was developed for methane cracking [8]. Its 10-kW$_{th}$ prototype was built and tested using natural gas, and 97% conversion was obtained with a maximum temperature above 2000 K under concentrated solar irradiation of 4 MW/m^2 [23]. The difficulties that prevent this technology to become commercial are mainly the cost of the reactor and the complicated flow pattern inside the reactors. For example, in order to prevent particle accumulation on the window, some inert gas is introduced to the reactor with high flow rates and pressures, or indirectly heated solar reactors are used which decreases the solar-to-fuel conversion efficiency and further increase the cost.

2.4. Gasification and Reforming of Coal and Biomass

Gasification is a chemical process that converts carbonaceous feedstock into gaseous fuels under a controlled amount of oxygen and/or steam [24]. Main difference between gasification and combustion is that products in gasification have useful heating value. In gasification, pressure inside the gasifier is generally in the range from 20 to 40 bar, whereas methanol or ammonia synthesis requires 50 to 200 bar [25]. In addition, temperatures inside the gasifier is generally in the range from 1400 to 1700°C [25].

Pyrolysis is a thermochemical process that occurs before gasification, and it decomposes the complex hydrocarbons into smaller and less complex molecules in the absence of oxidizers. In pyrolysis, the yield of solar char can be maximized by slowing the heating rate, lowering the temperature or allowing a longer residence time [26]. On the contrary, a higher heating rate, a higher temperature, and a shorter residence time maximize the gas yield. Additionally, liquid yield at an intermediate temperature can be maximized by increasing the heating rate or minimizing the residence time. Tar is an undesired by-product of gasification and pyrolysis. It can cause condensation and consequent plugging, formation of aerosols and polymerization into more complex structures [26].

Gasification is an endothermic process and requires energy to occur. In case of conventional gasification, this energy is supplied from the partial combustion or gasification of feedstock which emits CO_2 to the atmosphere. Use of concentrated solar energy eliminates or reduces the CO_2 emission and utilizes the clean high-temperature gasification process. Additionally, fuel value of the feedstock is increased with solar gasification. For example, fuel value of coal can be increased by about 45% using solar coal gasification [27], and CO_2 emission can be reduced by about 30% when compared to conventional coal gasification [28].

Solar gasification of coal and other carbonecous products is the process of converting these feedstock materials into some synthesis gas (syngas) which includes H_2, CO, CO_2 and water vapor using solar energy [29]. The gasification products can be further processed. For example, syngas can be processed to form methanol or ammonia or used in cement production, and lean gas can be combusted for heating or used in power stations to generate electricity [16, 30]. Solar gasification can be performed using CO_2or steam. In general, steam gasification of coal can be written as,

$$\text{Coal}+aH_2O \rightarrow bH_2+cCO+dCO_2 \tag{8}$$

This process is endothermic and requires temperatures above 1000°C. Solar gasification of petcoke, coal and other carbonecous feedstock started with directly irradiated solar reactors [31]. These designs have high reaction rates and kinetic and high fuel-to-product conversion. However, these reactors have problems with their aperture cover. As a cover, quartz window is commonly used to allow the concentrated solar power into the reactor. In directly irradiated reactors, quartz window has to withstand the high pressures inside the reactor and should not be covered with particles as the gasification occurs. As in the solar cracking, additional flows of inert gases are introduced into the reactor to prevent particle accumulation on the quartz window, but these additional flows introduce additional complexity and cost to the reactor [30]. In EU project SOLSYN [31], a 5-kW reactor prototype was built for solar coal gasification. The temperatures in this reactor could go up to 1700°C with the solar concentrating ratio of nearly 3000, but the general operation temperature was kept at 1220°C. The peak conversion efficiency was found to be 29% [31].

Similar to coal and other carbonecous feedstock, biomass can also be gasified in solar reactors. Conventionally, gasification of biomass has been done using the exhaust gas of combustion of fossil fuels or biomass itself. Biomass includes demol wood, wood chips, sewage sludge, almond shells, straw, etc. If the biomass is used, nearly 30% of the initial biomass has to be combusted with oxygen to drive the gasification process due to the temperature requirement [15]. This temperature requirement varies between 600 – 1000°C [32]. Additionally, one of the other disadvantages of conventional biomass gasification is the formation of tar which blocks and clogs the equipment. There have been some efforts to eliminate the tar formation with proper selection of materials, operating conditions and the design of the gasifier [32].

Solar-assisted gasification of biomass has advantages over the conventional process. The main advantages are the elimination of tar formation, even at temperatures as high as 1200°C, and high and rapid conversion of biomass to syngas. At the National Renewable Energy Laboratory (NREL) of US, bluegrass was gasified with a maximum conversion of 95% and about 5% of the products were hydrocarbons, ash and char [33]. The resident times can be less than 5 seconds [33]. There is also a solar reactor design to combine solar biomass gasification and steam reformation [33].

3. CONCENTRATED SOLAR REACTORS

This section defines and compares different solar concentrators and gives examples of directly irradiated and indirectly heated solar reactors for the solar fuel production processes defined in the previous section.

3.1. Solar Concentrators

There are two main types of concentrated solar collectors, categorized depending on their optical configurations. First type is parabolic trough systems in which there is an absorber tube in the focal line of parabolic reflectors. Linear Fresnel reflectors can also be included in this type of concentrated solar collectors. Second type is point focus solar collectors which include dish systems and heliostats. Dish systems have a solar receiver located in the focal point of the paraboloidal concentrator, and heliostats direct sun light to a solar receiver located at the top of a solar tower. Figure 3 shows the schematic of each solar collector type. Before going into details of each collector type, some terms need to be defined.

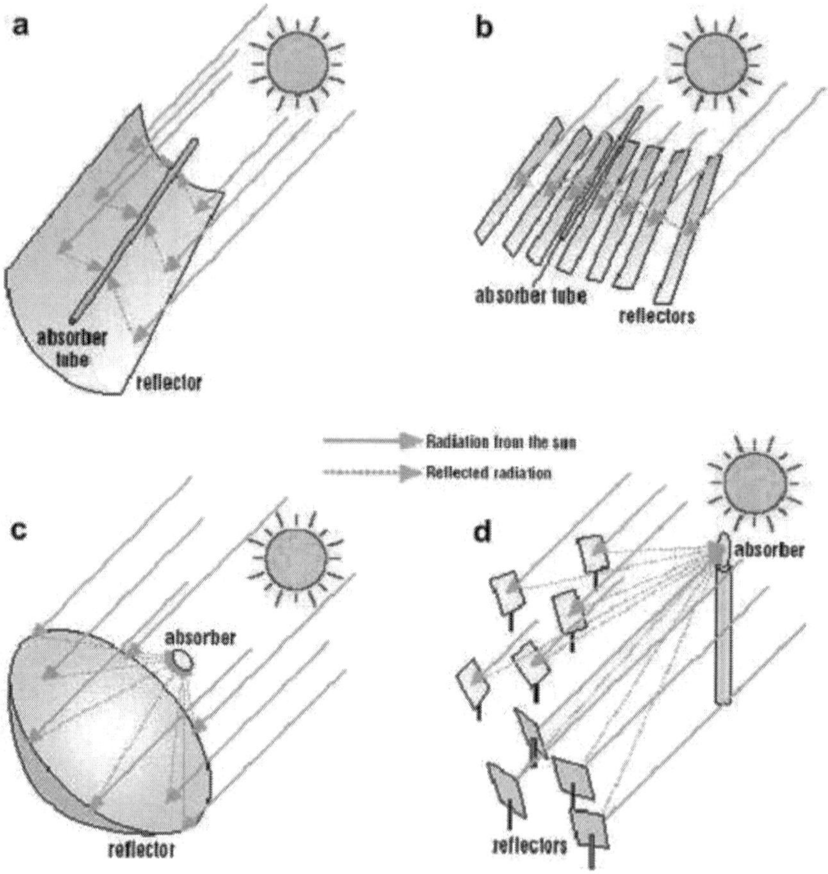

FIGURE 3. Solar concentrators, (a) parabolic trough, (b) linear Fresnel, (c) dish collectors, and (d) heliostats with solar tower [34].

Concentrating ratio is the ratio of the radiation intensity at the receiver by the radiation intensity received by the concentrator. Concentrating ratio plays an important role on the maximum achievable temperature at the receiver. Solar collector efficiency, η, is a product of Carnot efficiency, $\eta Carnot$, and the receiver efficiency, $\eta_{receiver}$, as,

$$\eta = \eta_{Carnot} \cdot \eta_{receiver} = \left(1 - \frac{T_o}{T_{rec}}\right)\left(\frac{\alpha \cdot G \cdot C - \varepsilon \cdot \sigma \cdot T_{rec}^4}{G \cdot C}\right) \quad (9)$$

where T_o and T_{rec} are the surroundings and receiver temperatures, respectively, α and ε are the absorptivity and emissivity of the receiver, G is the solar irradiation, C is the concentrating factor and σ is the Stefan-Boltzmann constant (5.67×10^{-8} W/m^2·K^4). Figure 4 shows the solar collector efficiency as a function of receiver temperature when ambient temperature is 300 K, absorptivity and emissivity are both 1, and the solar irradiation is 1000 W/m^2. The figure shows that higher thermal efficiencies and higher receiver temperatures can be obtained with increasing concentrating ratios. Therefore, the selection of solar concentrators mainly depends on the temperature requirement of the application.

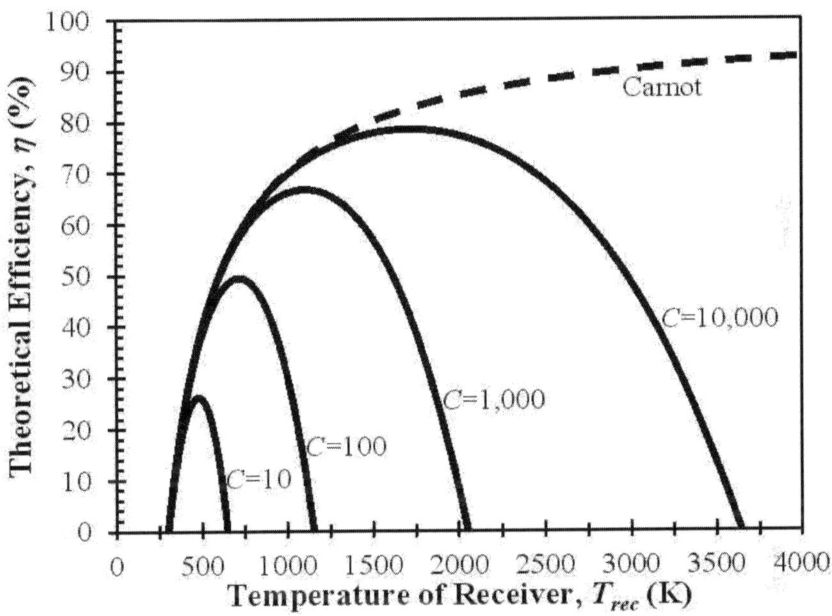

FIGURE 4. Theoretical solar reactor efficiency, η, as a function of receiver temperature, *Trec*, for different concentrating ratios, *C*.

3.1.1. Parabolic-Trough Systems

Parabolic-trough collectors consist of several tubes interconnected in the focal line of highly reflective sheet material. These linearly connected tubes are generally referred as evacuated tubes since they consist of two concentric tubes whose annulus is vacuumed. The working fluid is circulated in the inner tube. The inner tubes are generally made of metals, and outside of the inner tube has selective coating to increase the absorption and decrease the heat loss. The outer tubes are generally made of glass, and they can also have selective coatings.

Applications of parabolic-trough collectors can be divided into two sections depending on the temperature of application. The low temperature applications, varies from 100°C to 250°C, include domestic hot water, space heating and heat-driven refrigeration [35]. Their concentrating ratios are between 15 and 20. The high temperature applications have temperatures up to 400°C, concentrating ratios of 20 to 30 [35]. Theoretical limit for concentrating ratio can go up to 100 [2]. These collectors are mainly used in power plants that are driven by steam. In the parabolic trough collectors, the pressure within the inner tube can reach 10 MPa.

There are other configurations of cylindrical absorber tubes which are not commonly used [36-38]. One type has a circulation tube inside the inner cylinder which carries the working fluid. This circulation tube is attached to the inner tube with a cylindrical fin. In this type, the absorbing surface is the fin itself, not the inner tube. Another type cylindrical absorber has a delivery tube inside the inner tube, and the working fluid delivered by this delivery tube fills the space in the inner tube. In another tube type, working fluid fills the annulus [37].

These parabolic-trough collectors and other cylindrical absorber tubes may have diffuse or specular reflectors at their back. Diffuse reflectors are generally flat surfaces that cover the entire back of arrays of tubular collectors. However, specular reflectors have parabolic surfaces, and they cover the back of only one cylindrical absorber. If specular reflectors are used, the absorber tubes have to be in the reflector's focal line.

Line focus collectors are mounted with axes either in north-south or east-west direction. Hence, single axis tracking for this kind of collectors is sufficient to track the sun throughout the year. Collectors with axes in north-south direction track the sun from sunrise to sunset each day. Alternatively, collectors with axes in east-west direction track the sun seasonally. The spacing between each line of collectors in a solar farm with parabolic trough collectors is determined considering sun shadow line in the winter when the solar radiation comes with a larger azimuthal angle [38].

As an example, Nevada Solar One is in operation in Boulder City, Nevada, USA since 2007, covers 400 acres and has a capacity of 64 MW [39]. Figure 5 shows a picture of this power plant. Another and largest power plant with parabolic trough collectors is Solar Energy Generating Systems (SEGS) VIII-IX, built on approximately 240 acres and operating at 80 MW each near Harper Lake, California, USA. SEGS are also integrated with conventional natural gas turbines to operate at nights. It was reported that solar energy covers about 90% of the power production [40].

Another design is the absorber tubes with Fresnel reflectors. Fresnel reflectors that are mounted close to the ground direct and concentrate solar irradiation to the absorber tubes that are elevated at a higher level than reflectors. A secondary reflector on the back side of the absorber tube is also use to direct all the irradiation to the absorber tube. The main advantages of using Fresnel reflectors are that they have less wind load than the reflectors of parabolic trough collectors since these reflectors are located at a lower position, and no requirement for vacuum in the absorber tubes and for rotating joints [42]. However, the main disadvantage of Fresnel reflectors is that they have lower concentrating ratios than parabolic trough collectors [43].

FIGURE 5. Parabolic-trough collectors in Nevada Solar One power plant [41].

In Calasparra, Spain, Novatec Biosol built a power plant with 28 rows of linear Fresnel reflectors that produces 30 MW of electrical power in an area of nearly 200 acres. Figure 6 shows a picture of this power plant. The power plant uses steam, and the temperature and pressure of the steam produced reach to 270°C and 55 bars, respectively [44, 45].

FIGURE 6. Linear Fresnel reflectors in the power plant Thermosolar Power Plant (PE2) in Spain [46].

3.1.2. Point Focus Collectors

Parabolic dish collectors concentrate sunlight to the focal point of the parabolic reflectors. These collectors have two-axis sun tracking system. In the focal point of the parabolic reflector, a working fluid is heated directly to a maximum temperature of about 1000 K [47]. This working fluid is generally used to drive a Stirling engine or a gas turbine to produce electricity. The typical parabolic dish collectors have a diameter of 5 to 10 m, and each can produce up to 0.4 MW [47, 48]. The concentrating ratios parabolic dish collectors vary between 1,000 and 10,000 [2]. The reflector is usually made of silver or aluminum coated glass. This kind of collectors can be used in applications with relatively low power requirement in remote areas.

One of the first examples of power plant using parabolic dish collectors was supposed to be Maricopa Solar Plant in Arizona, USA before its contractor company was announced bankruptcy in 2011. Figure 7 shows a picture of this power plant. This power plant consists of 60 parabolic dish collectors that heat the hydrogen to drive Stirling engines. The power plant has a capacity of 1.5 MW. This technology is not commercially operational and available in large scale power production.

FIGURE 7. Parabolic dish collectors in Arizona, USA [49].

Some solar thermal power plants use arrays of heliostats which are sun-tracking flat mirrors. These mirrors or heliostats with two-axis tracking system direct solar irradiation to the receiver located at the top of a tower on a concrete support. In order to direct the sunlight to the receiver with sufficient accuracy all the times, a motor drive system with a large gear reduction is necessary [50]. Due to the presence of the tower, these power plants are sometimes referred as power tower systems. Concentrating ratios for these systems vary between 500 and 5000 [2], and the temperature at the receiver can exceed 2000 K depending on the concentrating ratio. These power plants can be used for converting solar energy to chemical energy, such applications include reduction of zinc oxide and coal gasification [51].

Throughout the technological development of heliostats, their sizes become larger and larger in order to decrease the production cost since their cost is a strong function of production rate. Although the initial development of heliostats started in 1975, one of the first prototypes of heliostats in 1980s by Sandia Labs, USA had an area of 37 m^2 [52]. Currently, Planta Solar (PS) 20 solar power plants use heliostats with each of their area as 120 m^2 [53].

Another improvement in the development of heliostats is the material choice. Glass mirrors with steel support structure are being replaced by silver polymer mirrors with silver-steel alloy structure in order to increase the structural durability and reduce the weight of heliostats [52]. Some designs also have circular mirrors instead of rectangular ones to reduce stress on the support structure.

The Crescent Dunes Solar Energy Project in Tonopah, Nevada, USA will be an example to the central tower power plants once it is completed late 2013. It will be built on approximately 1600 acres, and it will produce 110 MW of electrical power using molted salt as a phase changing storage medium [54]. As another example, PS 20 which is operational since 2009 has a cavity receiver at the top of a 165-m tower. Figure 8 shows the picture of this power plant. It heats up water in the cavity, and steam reaches an outlet temperature of maximum 550 K. This solar power plant consists of 1255 heliostats with a total area of 30 acres. The solar power plant is backed up with natural gas burnt conventional turbine, and the total power production is 20 MW. PS 20 power tower is cooled with water which is generally replaced by air cooling if the power plant is built on deserts due to lack of water resources. The cooling is necessary for the materials used in the power towers.

FIGURE 8. PS 10 (back) and PS 20 (front) solar thermal power plants with heliostats with solar towers [55].

To reach the necessary temperatures for the solar fuel production methods given in the previous section, tower or dish type collectors should be used. For further reference, a review paper on the volumetric receivers for the concentrating solar thermal power plants discussed different designs from the projects of the last 3 decades [56]. Another good review was done on comparison of parabolic trough, dish systems, solar towers and tubular systems with Fresnel reflectors by Pavlovic *et al.* [57].

3.2. Solar Reactors

In this section, different solar reactors that were designed for fuel production using concentrated solar energy are discussed and compared in terms of their operating conditions and design parameters. Depending on the reactor design, these reactors were fed feedstock with or without solid particles to perform hydrolysis, cracking, gasification, etc. These particles not only allowed a more uniform temperature distribution inside the reactor, but also helped the reactor to reach higher temperatures faster. These particles also acted as additional reaction sites due to their high surface are to volume ratios. In some designs, feedstock was diluted with some inert gas, such as Argon, to increase the produced fuel yield. An auxiliary gas was also fed to prevent particle deposition on the window surface. In some designs, reactors were supplemented with a cooling system for products to prevent them recombine.

In general, the solar energy is transformed into thermal energy in the structure of volumetric receivers. In some designs, solar energy directly heats the feedstock in the reactor which is referred as directly irradiated solar reactors. Additionally, some of these receivers have a porous metal or ceramic absorber to be heated by solar energy. Metal absorbers can be heated up to 1000°C whereas SiC absorbers can reach 1500°C as maximum temperatures. Then, this thermal energy is transferred to a working fluid that passes through the porous absorber. This kind of reactors is referred as indirectly heated solar reactors.

3.2.1. Directly Irradiated Solar Reactors

In this section, examples of directly irradiated solar reactors are presented with their design parameters, temperature and pressure allowances, their power outputs and their solar fuel production rates. These examples are selected to give a wide range of applications and designs.

Maag *et al.* [17] tested a 5 kW-prototype of a solar reactor seeded with particles for thermal cracking of methane. The cylindrical reactor was 200 mm in length and 100 mm in diameter. It had a 60 mm aperture area. The concentrator was covered with a 240 mm-diameter quartz window as shown inFigure 9. In their experiments, they used a sun-tracking parabolic concentrator that has a diameter of 8.5 m and could reach a concentrating factor up to 5000 suns. They tested the reactor in the temperature range from 1300 to 1600 K with a concentrating factor of 1720 suns. They varied the volume fraction of carbon in the range of 0 to $7.2x10^{-5}$ and gas inlet flow rate in the range of 8.6 to 15.6 l/min. They reported maximum methane-to-hydrogen conversion of 95% at a residence time less than 2 seconds, and an experimental solar-to-chemical energy conversion efficiency of 16%, whereas their theoretical prediction of the same conversion efficiency was 31%.

Yeheskela and Epstein [58] developed and tested 10-kW particle-seeded solar chemical reactor for producing hydrogen and carbon nanotubes from methane. They used iron pentacarbonyl and ferrocene as catalysts to produce multi-walled carbon nanotubes. The reactor was 300 mm in length, and the quartz window which covered the reactor as shown in Figure 10 was 200 mm in diameter. Additionally, He was used as a screen protector gas to eliminate the particle deposition on and near the window, and N_2 was used as a tornado generator gas. The average reported temperature within the reactor core was 1450°C.

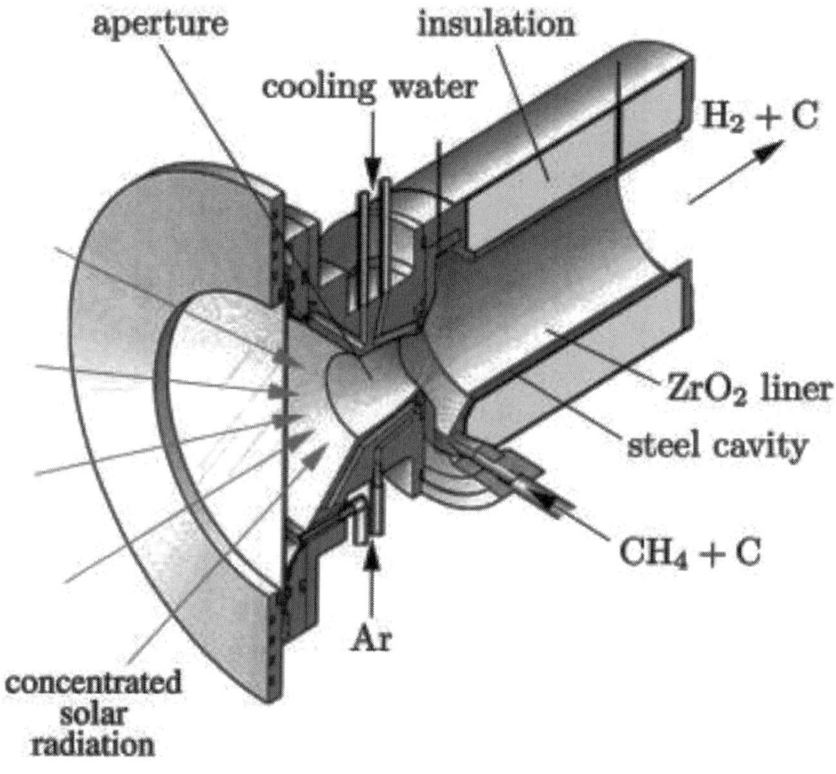

FIGURE 9. Schematic of design of Maag *et al.* [17].

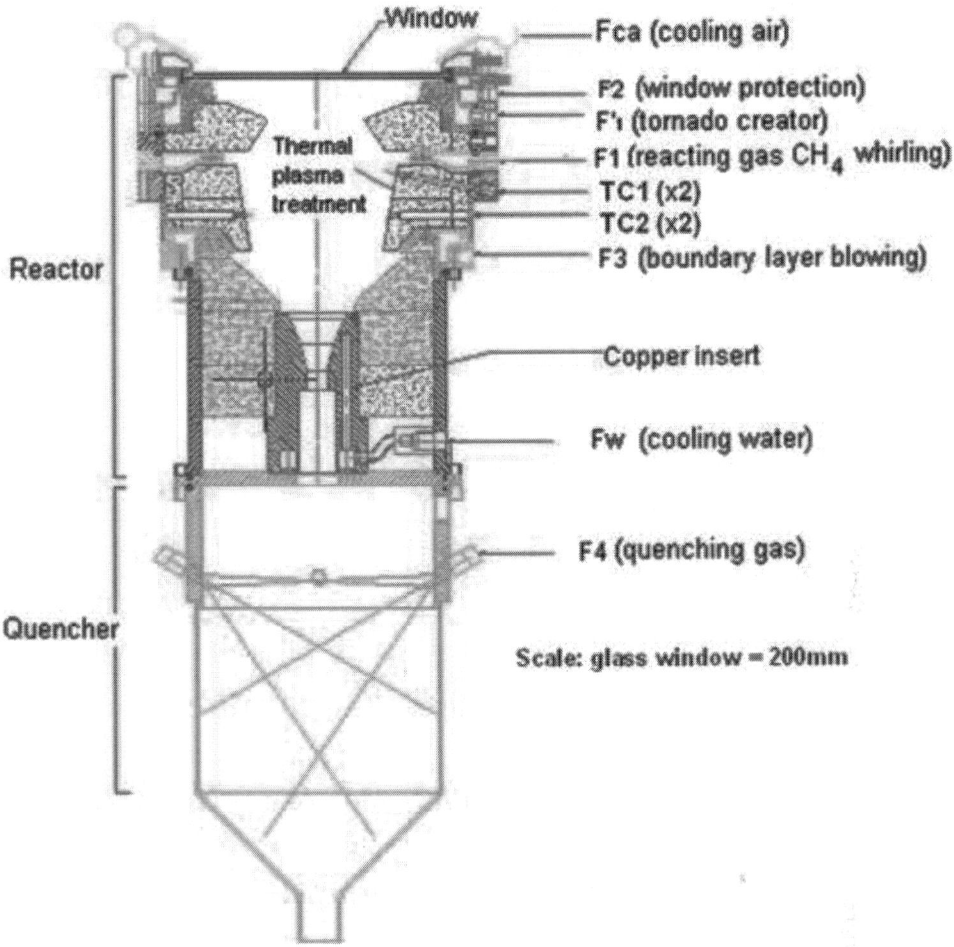

FIGURE 10. Schematic of design of Yeheskela and Epstein [58].

Abanades and Flamant [59] designed nozzle-type directly-irradiated solar reactor for methane cracking. They used a graphite nozzle with an inner diameter of 10 mm and a length of 65 mm. The schematic of the proposed design is shown in Figure 11. The reactor walls were made of stainless steel, and they were water cooled for their durability. Additionally, the products were also cooled to eliminate recombination of products [60]. The obtained conversion of methane to hydrogen exceeded 95% in molar basis, while the rest of the by-products were C_2H_2, C_2H_4 and C_2H_6. With a direct normal irradiation of 980 W/m^2, the temperature of graphite nozzle had a maximum temperature of 1385°C, while their model estimated the maximum wall temperature as 1890 K.

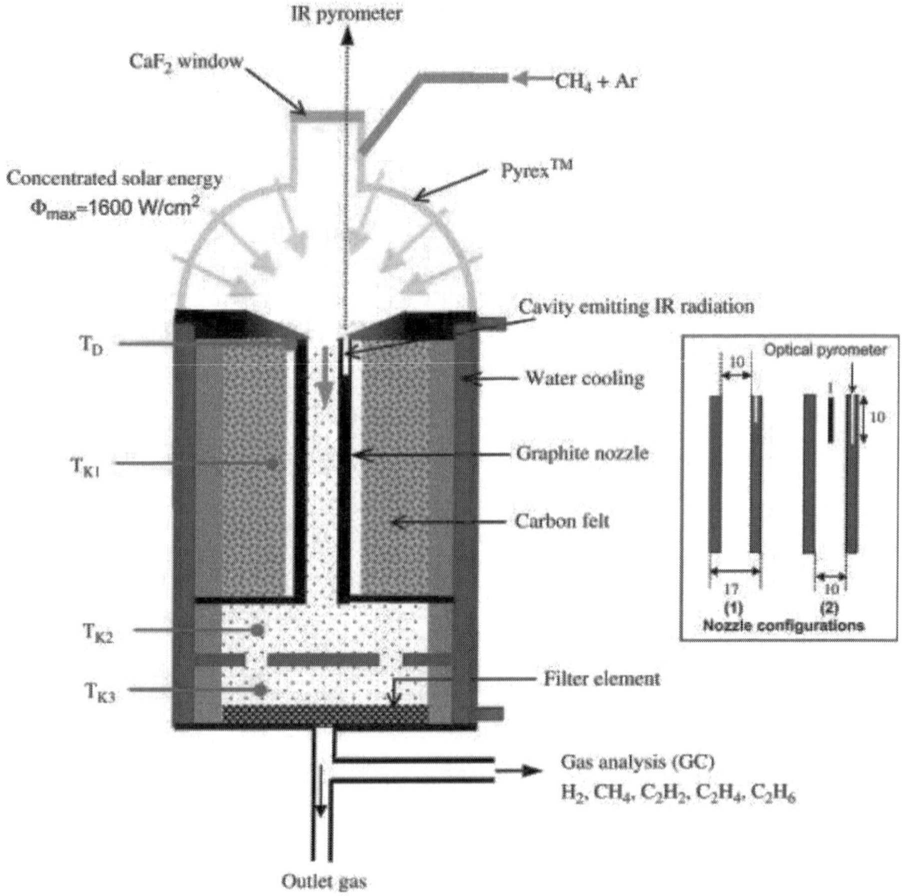

FIGURE 11. Schematic of design of Abanades and Flamant [59].

Klein *et al.* [61] investigated the performance of directly irradiated solar methane cracking process with and without CO_2 present in the reactor. The reactor, as shown in Figure 12, had a diameter of 160 mm, a length of 266 mm and an aperture diameter of 60 mm. The gas exit temperatures with CO_2 only (no methane) in the reactor were in the range from 1000 to 1250°C. Additionally, the exit temperatures were in the range from 1100 to 1450°C with CO_2 and methane, when the CO_2 and methane molar ratio were varied from 1:1 to 1:6. Overall, the experimental results were similar to the studies where no $CO_2$was introduced into the reactor. Moreover, when the reactor was fed with CO_2 and carbon black and the gas exit temperature reached 1000°C, 20% of carbon particles reacted with CO_2. When the exit gas temperature was increased to 1350°C, about 70% of the carbon particles reacted with CO_2.

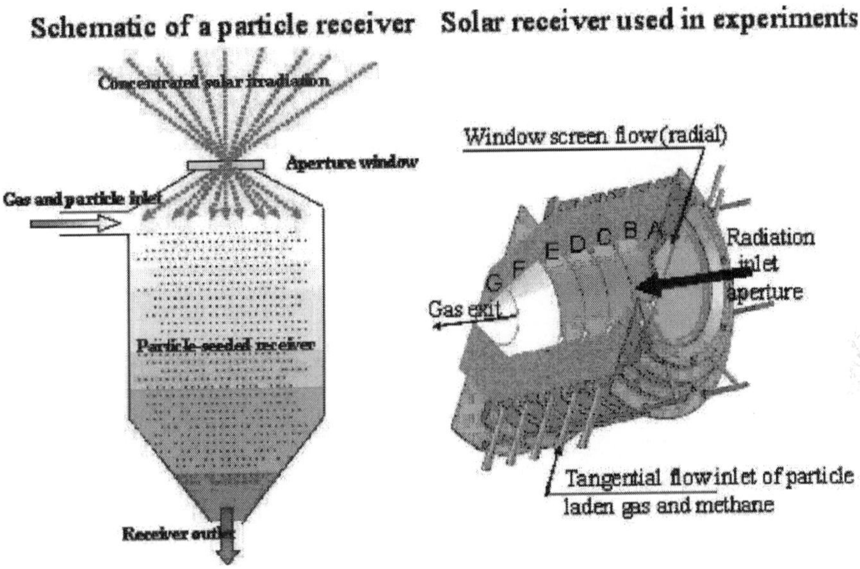

FIGURE 12. Schematic of design of Klein *et al.* [61].

Z'Graggen *et al.* [62] designed a 5-kW prototype reactor for steam-gasification of petroleum coke using concentrated solar energy. The reactor had a 5-cm diameter aperture which was covered by 3-mm-thick quartz window. The window was cooled by oil, and swept by an inert gas to prevent particle accumulation on the window. The solar concentrating ratio was about 5000, and the maximum temperature in the reactor was about 1800 K. The walls of the cavity were covered with Al_2O_3 and insulated from the backside with Al_2O_3-ZrO_2. Figure 13 shows the design of Z'Graggen *et al.* [62]. As a result of the steam-gasification of petroleum coke, H_2, CO, CO_2 and CH_4 were produced with a chemical conversion ratio of 87%. The overall solar-to-chemical conversion efficiency was about 9%.

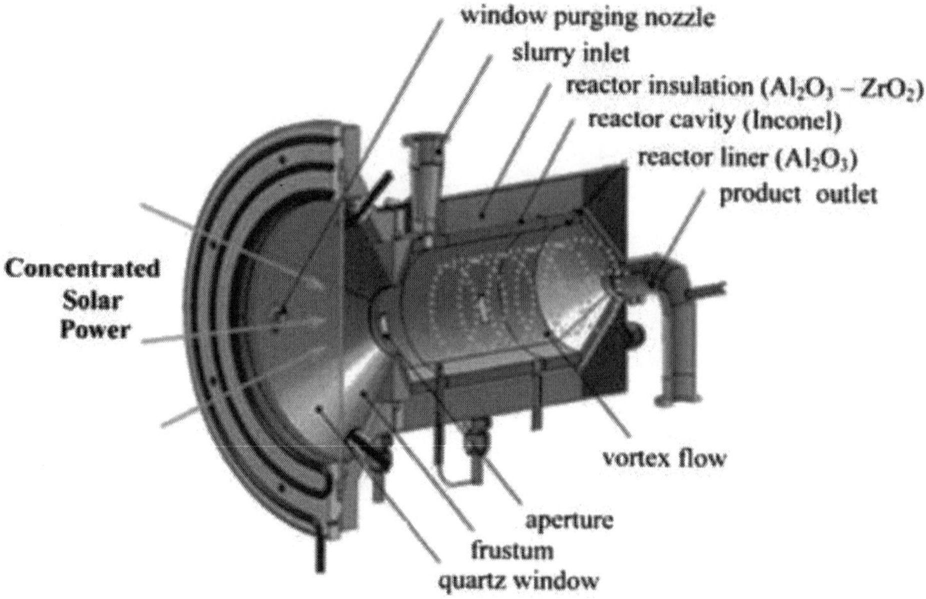

FIGURE 13. Reactor design of Z'Graggen *et al.* [62].

3.2.2. Indirectly Heated Solar Reactors

Directly irradiated solar reactors work with high conversion efficiencies. However, they have problems, such as accumulation of particles on the window. In order to overcome this problem and the need for inert gas feeding, indirectly heated solar reactors are alternatively introduced. This section gives some design examples of indirectly heated solar reactors for different solar fuel production methods.

Gordillo and Belghit [63] modeled the reaction kinetics without pyrolysis using finite volume analysis in a two-phase biochar solar gasification reactor with a bubbling fluidized bed as shown in Figure 14. Bubbling was used to introduce fluidizing gases to the reactor. They found that concentrating solar energy and high gas flow rates affected the temperature distribution within the packed bed. Therefore, a uniform temperature distribution could not be obtained which adversely affected the reaction kinetics. Additionally, they showed that although energy conversion efficiency (η), defined as in Equation (10), could be as high as 55%, it decreased with increasing the steam velocity or the bed temperature [64].

$$\eta = \frac{\dot{m}_{product} LHV_{product}}{\dot{Q}_{solar} + \dot{m}_{feedstock} LHV_{feedstock}} \quad (10)$$

where $\dot{m}$ and *LHV* refer to the mass flow rate and lower heating value, Q_{solar} is the solar irradiation, subscripts *product* and *feedstock* denote gaseous products and fed feedstock, respectively.

According to Hathaway *et al.* [65], problems with the preceding reactor designs had poor heat transfer characteristics [63], formation of ash and tar which block the radiative heat transfer and insulate the reaction zone, and intermittency of solar energy. Hathaway *et al.* [65] investigated the effects of using molten salt on the reaction kinetics in solar gasification of biomass. For the analysis of pyrolysis which occurs before gasification, they prepared tablets using microcrystalline cellulose, and for the analysis of steam gasification, they used tablets of wood charcoal powder. They carried out a series of experiments in the temperature range from 1100 to 1250 K to show the effects of molten alkali carbonate salts (lithium, sodium and potassium carbonate) on reaction rates using the experimental setup shown in Figure 15. They showed that introducing molten salts increased the rate of pyrolysis by 74% and increased the rate of gasification by more than an order of magnitude since molten salts acted as a heat transfer medium for gasification which ended up with more uniform temperature distribution within the solar reactor. On the contrary to the other studies, the catalytic effect of molten salt on pyrolysis was not observed for the reason that pyrolysis happened rapidly, and then gasification occurred. However, the catalytic effect of molten salt on gasification was observed. Introducing the molten salt increased the pre-exponential factor (i.e., rate of reaction in steady state process) by 24.4 times and increased the activation energy by about 4%. Additionally, using molten salt avoids the tar production (as a by-product of uncatalyzed gasification, occurs especially on startup), and molten salts can act as an energy storage unit to overcome the intermittency effect of solar energy.

Rodat *et al.* [66] developed a 10 kW tubular reactor prototype, which was indirectly heated, for methane cracking as shown in Figure 16. They used a graphite cubic cavity as a receiver and a quartz window. The quartz window was swept by nitrogen which prevented O_2 to enter the cavity. The reactor reached 2070 K, and the products included C_2H_2 with maximum mole fraction as 7%. As given in Equation (7), C_2H_2 is the last step of H_2 and carbon black evolution. The graphite cavity was purged by N_2. For this configuration, the reactor required about 4000 seconds to reach the required temperature of 1800 K when the experiment started at 300 K under the direct normal irradiance of 1000 W/m^2 [67].

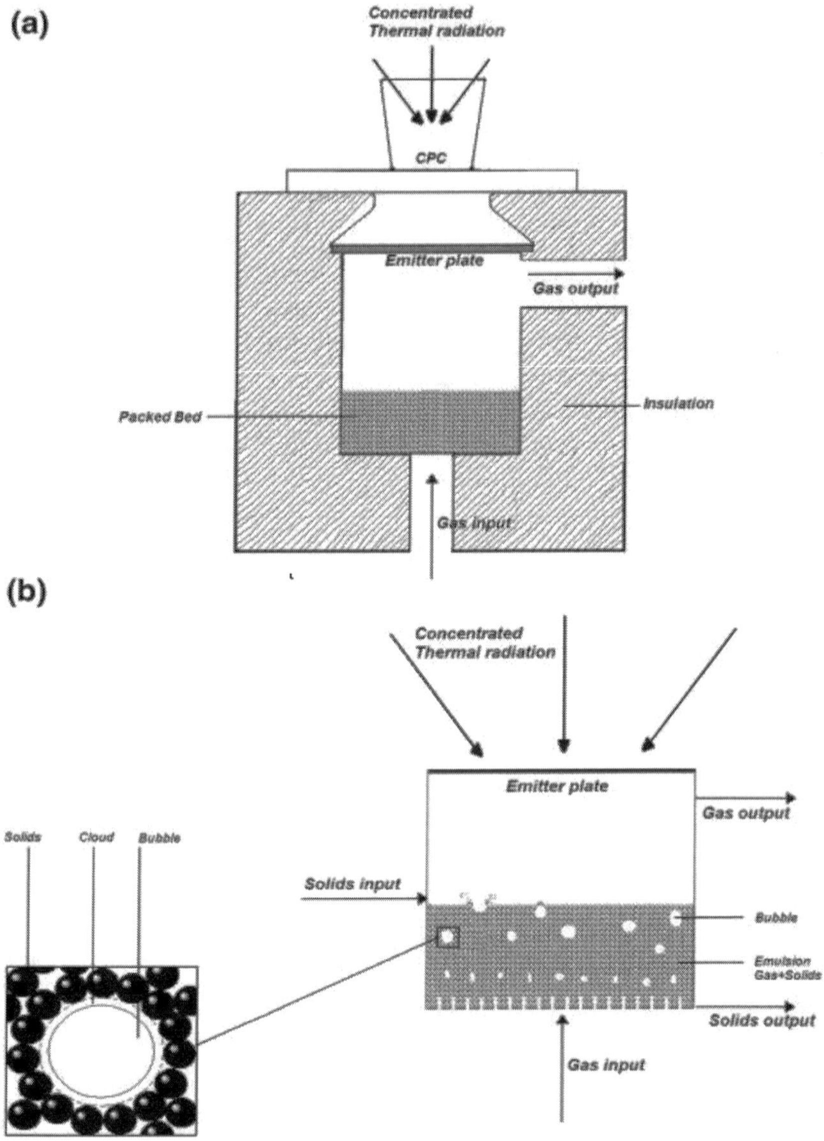

FIGURE 14. Model of Gordillo and Belghit [63].

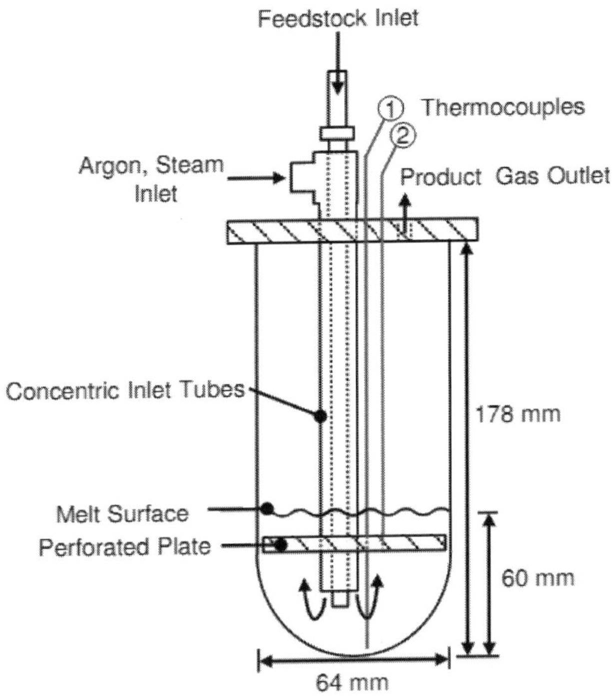

FIGURE 15. Experimental setup of Hathaway *et al.* [65].

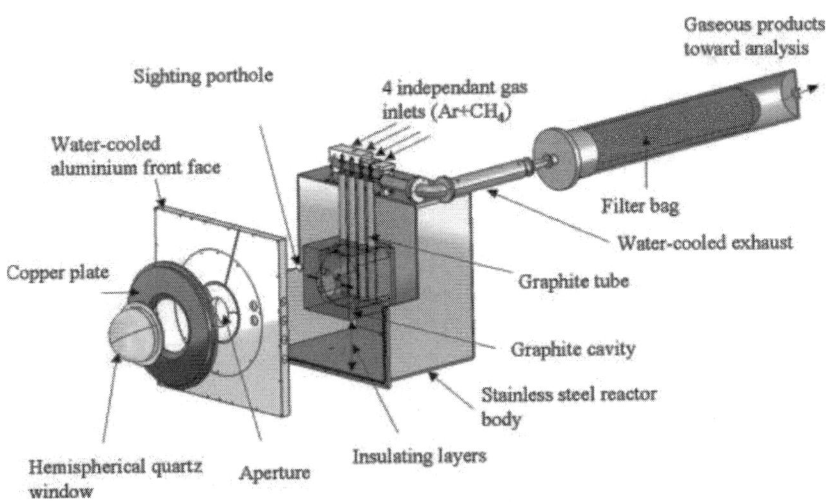

FIGURE 16. Schematic of design of Rodat *et al.* [66].

Lichty *et al.* [69] designed and analyzed the thermal characteristics of a cavity reactor prototype for solar-thermal biomass gasification as shown in Figure 17. The maximum recorded temperature was 1660 K on the central tube under 7.5 kW power input. They quantified the reacted biomass based on CO and CO_2 as these gases showed the ratio of reactants underwent a complete reaction, and the authors reported an average biomass-to-CO and CO_2 conversion as 58.4%. The residence time was about 4 seconds. They also compared the syngas production of grass and lignin pyrolysis and cellulose gasification using mass spectrophotometer.

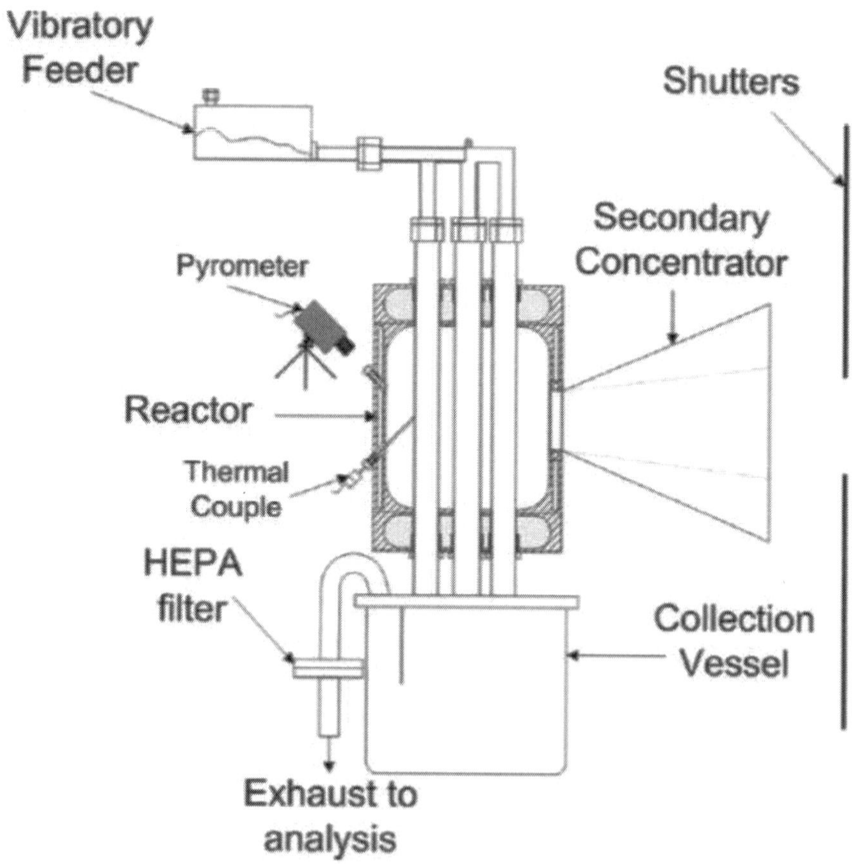

FIGURE 17. Reactor design of Lichty *et al.* [69].

In the design of the *German Aerospace Center* for directly irradiated solar reactor to reform natural gas is given in Figure 18, porous ceramic absorber coated with Rh catalyst was used [12]. A concave quartz window was also mounted on the concentrating solar collector [12]. The operating conditions were chosen as 1400°C and 3.5 bars, and a volumetric flow rate of 3.8 l/min with 5% methane in argon [70,71].

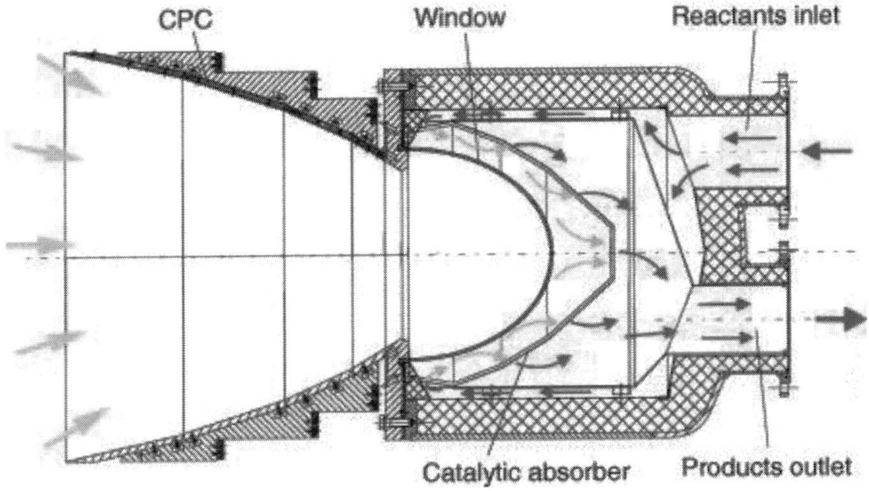

FIGURE 18. Design of the *German Aerospace Center* for natural gas reforming [12].

Maag *et al.* [72] simulated the performance of a 10 MW commercial-size reactor. The reactor consisted of four graphite absorber tubes with an outer diameter of 24 mm placed in a 0.2 m-cubic graphite cavity as shown in Figure 19. The graphite cavity had an aperture of diameter 9 mm which was covered by quartz window. They predicted 100% methane-to-hydrogen conversion when flow rate of methane was 0.7 kg/s at a reactor exit temperature of 1870 K. Spectral properties of quartz window were estimated using a band model, and view factors were calculated using Monte Carlo ray-tracing method. The energy balance for the overall system was solved with finite volume method. The results showed that it was possible to increase solar-to-chemical energy conversion efficiency from 42% to 60% when the outlet temperature was lowered to 1600 K and, subsequently, the methane flow rate was doubled, but then quality of carbon black as a product would be poorer.

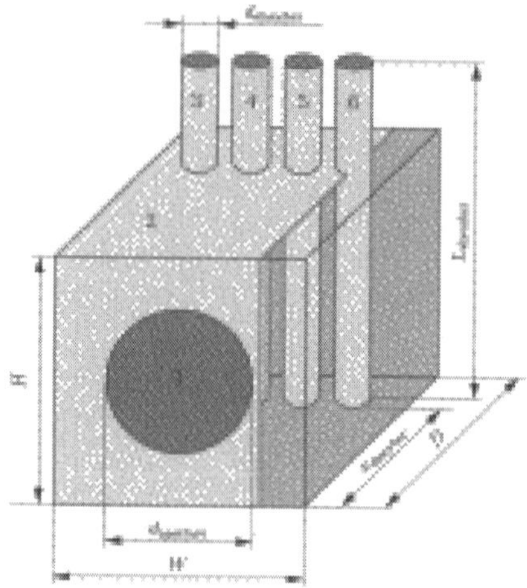

FIGURE 19. Schematic of design of Maag *et al.* [72].

Osinga *et al.* [73, 74] designed 5-kW indirectly heated solar reactor for the reduction of ZnO. There were two different versions of the reactor. First one had the inner cavity made of graphite, and the second one had the inner cavity made of SiC. Figure 20 shows the second type, reactor with SiC absorber. Both reactors could reach temperature of 1700 K in about 80 minutes after the solar energy was input to the reactor. The reactor with graphite absorber had a vacuum pressure of 10 mbar whereas the pressure inside the reactor with SiC absorber was kept at 1 bar. ZnO and C mixture was reduced to Zn, CO and CO_2 from which Zn can be reacted with water to produce ZnO and H_2 as in Equation (3)[75].

The reactor with SiC absorber was scaled up to a capacity of 300 kW and operated by Wieckert *et al.*[76] under the EU project, called SOLZINC. They used solar tower with heliostats to deliver concentrated solar power of 300 kW to the volumetric reactor. The second cavity where the mixture of ZnO and C was present reached to maximum temperature of nearly 1500 K, and overall ZnO-to-Zn conversion reached 95% with a Zn production of 50 kg/h.

Piatkowski *et al.* [31, 77] designed a 5-kW indirectly heated solar reactor with packed bed for solar steam-gasification. Figure 21 shows the reactor design of Piatkowski *et al.* The authors used different carbonecous feedstock, such as African coal, sludge and charcoal. Beech charcoal gave the maximum solar-to-chemical conversion efficiency of 29% at a temperature of nearly 1500 K. The solar reactor has two cavities separated an emitter plate made of SiC-coated graphite. The aperture diameter was

6.5 cm which was followed by a 3-mm-thick quartz window. The walls of the lower cavity was also covered by SiC, and insulated by Al_2O_3-SiO_2. The reactor received the concentrated solar energy with concentrating ratios up to 3000, and the maximum temperature measured in the upper cavity was about 1700°C.

Summary of the operating conditions of the discussed designs for methane cracking is given in Table 3.

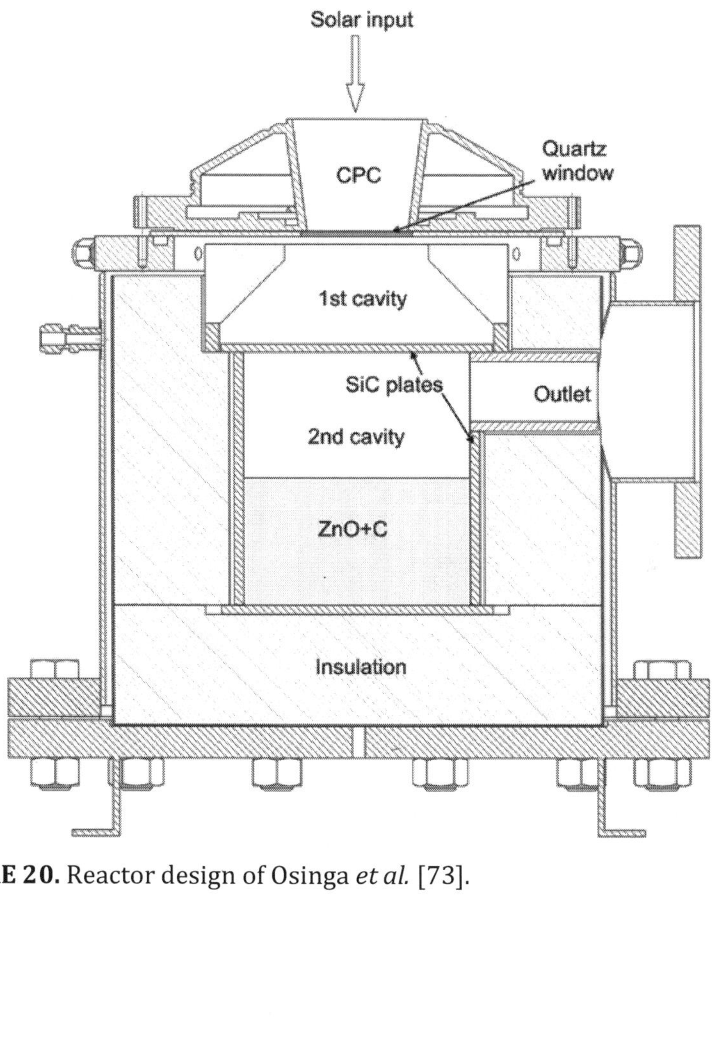

FIGURE 20. Reactor design of Osinga *et al.* [73].

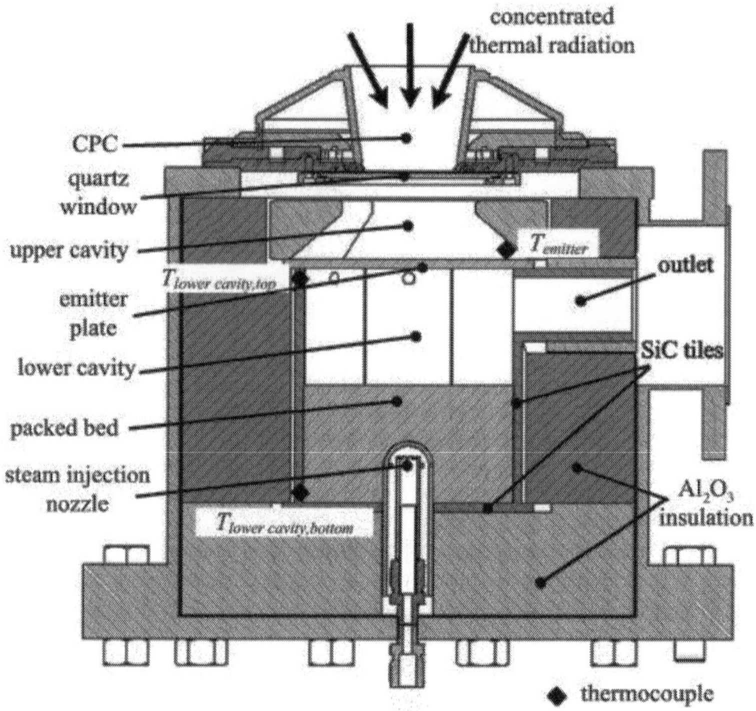

FIGURE 21. Reactor design of Piatkowski *et al.* [31, 77].

TABLE 3. Operating Conditions of Different Reactor Designs for Methane Cracking.

Reference	Maximum Temperature (°C)	Inlet CH_4 Dilution (%vol.)	Reactor Dimensions (mm)	Aperture Diameter (mm)	Inlet Flow Rate(l/min)	Catalytic or Fed Conversion
Directly Irradiated Solar Reactors						
Maag *et al.* [17]	1327	6-30 (in Argon)	100 (diameter) 200 (length)	60	8.6-15.6	Carbon black seeded
Yeheskela and Epstein [58]	1450	98 (in catalysts)	200 (diameter) 300 (length)	200	5-9.7	Flow with $Fe(CO)_5$, $Fe(C_5H_5)_2$
Abanades and Flamant [59, 60]	1110	11-20(in Argon)	10 (diameter) 65 (length)	10	0.9	No particle feeding
Klein *et al.* [61]	1471	10-24 (in Argon or CO_2)	160 (diameter) 266 (length)	60	37-60	Carbon black seeded

Indirectly Heated Solar Reactors						
Rodat *et al.* [66]	1800	10-20 (in Argon)	18 (tube diameter) 200 (cube side)	90	-	No particle feeding
German Aerospace Center [12]	1400	5 (in Argon)	-	-	3.8	Reactor walls with Rh
Maag *et al.* [72]	1600	10-20 (in Argon)	24 (tube diameter) 200 (cube side)	9	10-48	No particle feeding

TABLE 4. Targets of Department of Energy of US [14] and predictions [78] for cost and efficiency for ZnO/Zn thermochemical cycle.

	by 2015		by 2025	
	DOE Target	Prediction‡	DOE Target	Prediction‡
Cost ($/gge†)	6	6.07	3	4.18
Process Efficiency (%)	30	35	35	42

†gge refers to gallon-of-gas-equivalent.

‡Based on predicted ZnO-to-Zn conversions of 70% in 2015 and 85% in 2025.

4. CONCLUSIONS AND OUTLOOK

The research to find an alternative fuel to fossil fuels is led by how the new technologies are economically competitive with the fossil fuel technologies, rather than their efficiencies. However, the economical aspect of fossil fuels should also include the cost for CO_2 emissions or sequestration of $CO_2$when fossil fuels are compared to solar fuels since solar fuels have no CO_2 emission. Furthermore, as the fossil fuels deplete and the demand for fossil fuels will exceed their production, their prices will be subjected to significant increase. In this way, the investors in fuel or electricity production sector would see solar fuels as an alternative to fossil fuels with the current level of both technologies. As an outcome, the world would become a more sustainable environment with reduced atmospheric CO_2 level and less pronounced risk for global warming.

It should also be noted that the solar fuel production methods introduced in this chapter are at different levels of maturity. For instance, most of the thermochemical cycles are in laboratory and research scale, whereas gasification and reforming processes are in fully operational or pilot stage. To give some examples on solar fuel production, the targets and predictions of the US Department of Energy (DOE) [78] for both cost and process efficiency are summarized in Table 4 for the ZnO/Zn thermochemical cycle. The predictions show that it is possible and feasible to meet the DOE efficiency and 2015 cost targets. However, the cost target of DOE in 2025 is a challenging objective. The main constituent of cost in thermochemical solar production is the plant capital cost, i.e., cost for heliostats and solar tower, rather than the direct cost for the process. Although process efficiencies are reported and predicted as given in Table 4, the overall solar-to-fuel conversions are still low, less than 10% [1].

In summary, the advantages of solar fuels include:

- Energy content or heating value of feedstock is increased by converting it to another form, solar fuel.
- Producing storable and transportable fuel which is not possible if solar energy is directly used. Thus, eliminates the intermittency problem of solar energy.
- Solar fuels are clean and sustainable. The thermochemical cycles and thermolysis of water that are used to produce solar fuels have no CO or CO_2 emissions. However, carbon emission occurs for the gasification or reforming of carbonecous feedstocks. If these feedstocks are biomass cultivated with CO_2 from the atmosphere, they are carbon neutral.

On the other hand, these are not mature technologies and still suffer from technical challenges which form the basis for future research including:

- High temperatures needed for solar fuel production processes. High temperatures can be reached with high concentrating ratios. However, high concentrating ratios bring high cost to the system, and high temperatures restrict the material choice.
- Recombination of product gases, especially in thermochemical cycles, is a significant problem. This recombination significantly decreases both the process and overall solar-to-fuel efficiency.
- Quenching is introduced to products in order to reduce the recombination. However, quenching adds additional cost and complexity to the reactor and the process management. For some solar thermochemical processes, *membranes* are also required to separate product gases.

- Particle accumulation on the window of the reactor is a problem in directly irradiated solar reactors. This problem can be eliminated by introducing an inert gas with high flow rates to the reactor which further complicates the management of reaction in the reactor. Another solution is to heat the reactor indirectly which reduces the solar-to-fuel efficiency.
- Multiple-step chemical reactions are needed to produce hydrogen in most of the thermochemical cycles. More reactions add further components to the system which increase the cost and the management of the overall fuel production process.
- CO and CO_2 formation can be noteworthy in case of solar gasification and reforming of carbonecous feedstock, although solar fuels are accepted as clean fuels.

These drawbacks of the solar fuel production prevent the technology to be converted to large scale commercially available power plants. However, solar fuel production processes are thermodynamically efficient, favorable developments to increase the feedstock's heating values with the unlimited free solar energy. Therefore, in a long-term prospect, solar fuel production is a promising technology that needs significant research efforts for efficiently producing clean and sustainable fuels.

REFERENCES

1. Nowotny J, Sorrell C, Sheppard L, Bak T. Solar-Hydrogen: Environmentally Safe Fuel for the Future. International Journal of Hydrogen Energy. 2005;30(5):521-44.
2. Steinfeld A, Palumbo R. Solar Thermochemical Process Technology. Encyclopedia of Physical Science and Technology. 2001;15(1):237-56.
3. Steinfeld A, Meier A. Solar Fuels and Materials. Amsterdam: Elsevier; 2004. p. 623–37.
4. Baykara S. Experimental Solar Water Thermolysis. International Journal of Hydrogen Energy. 2004;29(14):1459-69.
5. Steinfeld A. Solar Hydrogen Production via a Two-Step Water-Splitting Thermochemical Cycle Based on Zn/ZnO Redox Reactions. International Journal of Hydrogen Energy. 2002;27(6):611-9.
6. Funke HH, Diaz H, Liang X, Carney CS, Weimer AW, Li P. Hydrogen Generation by Hydrolysis of Zinc Powder Aerosol. International Journal of Hydrogen Energy. 2008;33(4):1127-34.

7. Abanades S, Charvin P, Lemont F, Flamant G. Novel Two-Step SnO2/SnO Water-Splitting Cycle for Solar Thermochemical Production of Hydrogen. International Journal of Hydrogen Energy. 2008;33(21):6021-30.
8. Meier A, Steinfeld A. Solar Thermochemical Production of Fuels. Advances in Science and Technology. 2011;74(1):303-12.
9. Chueh WC, Haile SM. Ceria as a Thermochemical Reaction Medium for Selectively Generating Syngas or Methane from H2O and CO2. ChemSusChem. 2009;2(8):735-9.
10. Kappauf T, Fletcher EA. Hydrogen and Sulfur from Hydrogen Sulfide—VI. Solar Thermolysis. Energy. 1989;14(8):443-9.
11. Zaman J, Chakma A. Production of Hydrogen and Sulfur from Hydrogen Sulfide. Fuel processing technology. 1995;41(2):159-98.
12. Steinfeld A. Solar Thermochemical Production of Hydrogen - a Review. Solar energy. 2005;78(5):603-15.
13. Harvey WS, Davidson JH, Fletcher EA. Thermolysis of Hydrogen Sulfide in the Temperature Range 1350-1600 K. Industrial & engineering chemistry research. 1998;37(6):2323-32.
14. Perret R. Solar Thermochemical Hydrogen Production Research (STCH), Thermochemical Cycle Selection and Investment Priority. Albuquerque, New Mexico and Livermore, California: Sandia National Laboratories 2011. Report No.: SAND2011–3622.
15. Perkins C, Weimer AW. Solar-Thermal Production of Renewable Hydrogen. AIChE Journal. 2009;55(2):286-93.
16. Zedtwitz P, Steinfeld A. The Solar Thermal Gasification of Coal—Energy Conversion Efficiency and CO2 Mitigation Potential. Energy. 2003;28(5):441-56.
17. Maag G, Zanganeh G, Steinfeld A. Solar Thermal Cracking of Methane in a Particle-Flow Reactor for the Co-Production of Hydrogen and Carbon. International Journal of Hydrogen Energy. 2009;34(18):7676-85.
18. Hirsch D, Epstein M, Steinfeld A. The Solar Thermal Decarbonization of Natural Gas. International Journal of Hydrogen Energy. 2001;26(10):1023-33.
19. Ozalp N, Kogan A, Epstein M. Solar Decomposition of Fossil Fuels as an Option for Sustainability. International Journal of Hydrogen Energy. 2009;34(2):710-20.

20. Rodat S, Abanades S, Coulié J, Flamant G. Kinetic Modelling of Methane Decomposition in a Tubular Solar Reactor. Chemical Engineering Journal. 2009;146(1):120-7.
21. Abanades S, Flamant G. Hydrogen Production from Solar Thermal Dissociation of Methane in a High-Temperature Fluid-Wall Chemical Reactor. Chemical Engineering and Processing: Process Intensification. 2008;47(3):490-8.
22. Almodaris M, Khorasani S, Abraham JJ, Ozalp N, editors. Simulation of Solar Thermo-Chemical Hydrogen Production Techniques. ASME/JSME 2011 8th Thermal Engineering Joint Conference; 2011 March 13-17, 2011; Honolulu, Hawaii, USA: ASME.
23. Rodat S, Abanades S, Flamant G. High-Temperature Solar Methane Dissociation in a Multitubular Cavity-Type Reactor in the Temperature Range 1823 – 2073 K. Energy & fuels. 2009;23(5):2666-74.
24. Kamka F, Jochmann A, Picard L, editors. Development Status of BGL Gasification. International Freiberg Conference on IGCC & XtL Technologies; 2005; Freiberg, Germany.
25. Higman C, Burgt Mvd. Gasification. Second ed. Oxford, UK: Gulf Professional Publishing; 2008.
26. Basu P. Biomass Gasification and Pyrolysis: Practical Design and Theory. Burlington, MA: Academic Press; 2010.
27. Kodama T, Kondoh Y, Tamagawa T, Funatoh A, Shimizu K, Kitayama Y. Fluidized Bed Coal Gasification with CO2 under Direct Irradiation with Concentrated Visible Light. Energy & fuels. 2002;16(5):1264-70.
28. Aoki A, Ohtake H, Shimizu T, Kitayama Y, Kodama T. Reactive Metal-Oxide Redox System for a Two-Step Thermochemical Conversion of Coal and Water to CO and H2. Energy. 2000;25(3):201-18.
29. Flechsenhar M, Sasse C. Solar Gasification of Biomass Using Oil Shale and Coal as Candidate Materials. Energy. 1995;20(8):803-10.
30. Van Heek K. General Aspects and Engineering Principles for Technical Application of Coal Gasification. In: Figuieiredo JL, Moulijn JA, editors. Carbon and Coal Gasification: Science and Technology. Alvor, Portugal: Springer; 1986. p. 383-402.
31. Piatkowski N, Wieckert C, Steinfeld A. Experimental Investigation of a Packed-Bed Solar Reactor for the Steam-Gasification of Carbonaceous Feedstocks. Fuel processing technology. 2009;90(3):360-6.

32. Devi L, Ptasinski KJ, Janssen FJJG. A Review of the Primary Measures for Tar Elimination in Biomass Gasification Processes. Biomass and Bioenergy. 2003;24(2):125-40.
33. Weimer A, Perkins C, Mejic D, Lichty P, inventors; WO Patent WO/2008/027,980, assignee. Rapid Solar-Thermal Conversion of Biomass to Syngas. United States patent WO2008027980. 2008 06.03.2008.
34. Quaschning V. Technology Fundamentals. Solar Thermal Power Plants. Renew Energy World. 2003;6(6):109-13.
35. Fernández-García A, Zarza E, Valenzuela L, Pérez M. Parabolic-Trough Solar Collectors and Their Applications. Renewable and Sustainable Energy Reviews. 2010;14(7):1695-721.
36. Morrison G, Budihardjo I, Behnia M. Water-in-Glass Evacuated Tube Solar Water Heaters. Solar energy. 2004;76(1):135-40.
37. Duffie JA, Beckman WA. Solar Engineering of Thermal Processes. 3rd ed. New Jersey: John Wiley & Sons; 2006.
38. Mills DR, Morrison GL. Compact Linear Fresnel Reflector Solar Thermal Powerplants. Solar energy. 2000;68(3):263-83.
39. Acciona North America. Nevada Solar One. http://www.acciona-na.com/About-Us/Our-Projects/U-S-/Nevada-Solar-One (accessed September 9, 2012).
40. California Energy Commission. Large Solar Energy Projects. http://www.energy.ca.gov/siting/solar/ (accessed September 9, 2012).
41. Wald ML. The New York Times: In the Desert, Harnessing the Power of the Sun by Capturing Heat Instead of Light. http://www.nytimes.com/2007/07/17/business/17thermal.html?_r=1 (accessed September 18, 2012).
42. Areva. Our Technology and Features. http://www.areva.com/EN/solar-198/arevasolarour-technology.html#tab=tab2 (accessed September 9, 2012).
43. Abbas R, Montes M, Piera M, Martínez-Val J. Solar Radiation Concentration Features in Linear Fresnel Reflector Arrays. Energy Conversion and Management. 2012;54(1):133-44.
44. National Renewable Energy Laboratory. Puerto Errado 2 Thermosolar Power Plant. http://www.nrel.gov/csp/solarpaces/project_detail.cfm/projectID=159 (accessed September 9, 2012).
45. Novatec Solar. Puerto Errado 2 in Spain. http://www.novatecsolar.com/56-1-PE-2.html (accessed September 9, 2012).

46. Tubosol PE2. Puerto Errado 2 (in German). http://www.puertoerrado2.com/ (accessed September 18, 2012).
47. Ummadisingu A, Soni M. Concentrating Solar Power–Technology, Potential and Policy in India. Renewable and Sustainable Energy Reviews. 2011;15(9):5169-75.
48. Kaygusuz K. Prospect of Concentrating Solar Power in Turkey: The Sustainable Future. Renewable and Sustainable Energy Reviews. 2011;15(1):808-14.
49. SRP. Tessera Solar and Stirling Energy Systems Unveil World's First Commercial-Scale Suncatcher Plant, Maricopa Solar – Peoria, Arizona. http://energydeals.wordpress.com/2010/06/04/tessera-solar-and-stirling-energy-systems-unveil-worlds-first-commercial-scale-suncatcher-plant-maricopa-solar-with-partner-srp/ (accessed September 18, 2012).
50. United States Environmental Protection Agency. Final Report: Design and Fabrication of a Reduced Cost Heliostat. http://cfpub.epa.gov/ncer_abstracts/index.cfm/fuseaction/display.abstractDetail/abstract/9033/report/F (accessed September 10, 2012).
51. Pitz-Paal R, Botero NB, Steinfeld A. Heliostat Field Layout Optimization for High-Temperature Solar Thermochemical Processing. Solar energy. 2011;85(2):334-43.
52. Jones SA, Lumia R, Davenport R, Thomas RC, Gorman D, Kolb GJ, et al. Heliostat Cost Reduction Study. Albuquerque, New Mexico and Livermore, California.: Sandia National Laboratories 2007. Report No.: SAND2007-3293.
53. National Renewable Energy Laboratory. Planta Solar 20. http://www.nrel.gov/csp/solarpaces/project_detail.cfm/projectID=39 (accessed September 10, 2012).
54. Solar Reserve. Crescent Dunes. http://www.solarreserve.com/what-we-do/csp-projects/crescent-dunes/ (accessed September 10, 2012).
55. Abengoa Solar. PS20, the Largest Solar Power Tower Worldwide. http://www.abengoasolar.com/corp/web/en/nuestras_plantas/plantas_en_operacion/espana/PS20_la_mayor_torre_comercial_del_mundo.html (accessed September 18, 2012).
56. Ávila-Marín AL. Volumetric Receivers in Solar Thermal Power Plants with Central Receiver System Technology: A Review. Solar energy. 2011;85(5):891-910.

57. Pavlović TM, Radonjić IS, Milosavljević DD, Pantić LS. A Review of Concentrating Solar Power Plants in the World and Their Potential Use in Serbia. Renewable and Sustainable Energy Reviews. 2012;16(6):3891-902.
58. Yeheskel J, Epstein M. Thermolysis of Methane in a Solar Reactor for Mass-Production of Hydrogen and Carbon Nano-Materials. Carbon. 2011;49(14):4695-703.
59. Abanades S, Flamant G. Experimental Study and Modeling of a High-Temperature Solar Chemical Reactor for Hydrogen Production from Methane Cracking. International Journal of Hydrogen Energy. 2007;32(10-11):1508-15.
60. Abanades S, Flamant G. Production of Hydrogen by Thermal Methane Splitting in a Nozzle-Type Laboratory-Scale Solar Reactor. International Journal of Hydrogen Energy. 2005;30(8):843-53.
61. Klein HH, Karni J, Rubin R. Dry Methane Reforming without a Metal Catalyst in a Directly Irradiated Solar Particle Reactor. Journal of solar energy engineering. 2009;131(2):021001.
62. Z'graggen A, Haueter P, Trommer D, Romero M, De Jesus J, Steinfeld A. Hydrogen Production by Steam-Gasification of Petroleum Coke Using Concentrated Solar Power—II Reactor Design, Testing, and Modeling. International Journal of Hydrogen Energy. 2006;31(6):797-811.
63. Gordillo E, Belghit A. A Bubbling Fluidized Bed Solar Reactor Model of Biomass Char High Temperature Steam-Only Gasification. Fuel processing technology. 2010;92(3):314-21.
64. Gordillo E, Belghit A. A Downdraft High Temperature Steam-Only Solar Gasifier of Biomass Char: A Modelling Study. Biomass and Bioenergy. 2011;35(5):2034-43.
65. Hathaway BJ, Davidson JH, Kittelson DB. Solar Gasification of Biomass: Kinetics of Pyrolysis and Steam Gasification in Molten Salt. Journal of solar energy engineering. 2011;133(2):021011.
66. Rodat S, Abanades S, Flamant G. Co-Production of Hydrogen and Carbon Black from Solar Thermal Methane Splitting in a Tubular Reactor Prototype. Solar energy. 2011;85(4):645-52.
67. Rodat S, Abanades S, Sans JL, Flamant G. Hydrogen Production from Solar Thermal Dissociation of Natural Gas: Development of a 10 kW Solar Chemical Reactor Prototype. Solar energy. 2009;83(9):1599-610.
68. Rodat S, Abanades S, Flamant G, editors. Hydrogen Production from Natural Gas Thermal Cracking: Design and Test of a Pilot-Scale Solar

Chemical Reactor. 18th World Hydrogen Energy Conference 2010; 2010 May 16-21, 2010; Essen, Germany.

69. Lichty P, Perkins C, Woodruff B, Bingham C, Weimer A. Rapid High Temperature Solar Thermal Biomass Gasification in a Prototype Cavity Reactor. Journal of solar energy engineering. 2010;132(1):011012.
70. Wullenkord M, Funken KH, Sattler C, Pitz-Paal R, Stolten D, Grube T, editors. Hydrogen Production by Thermal Cracking of Methane–Investigation of Reaction Conditions. 18th World Hydrogen Energy Conference; 2010; Essen, Germany.
71. Worner A, Tamme R. CO2 Reforming of Methane in a Solar Driven Volumetric Receiver-Reactor. Catalysis today. 1998;46(2-3):165-74.
72. Maag G, Rodat S, Flamant G, Steinfeld A. Heat Transfer Model and Scale-up of an Entrained-Flow Solar Reactor for the Thermal Decomposition of Methane. International Journal of Hydrogen Energy. 2010;35(24):13232-41.
73. Osinga T, Olalde G, Steinfeld A. Solar Carbothermal Reduction of ZnO: Shrinking Packed-Bed Reactor Modeling and Experimental Validation. Industrial & engineering chemistry research. 2004;43(25):7981-8.
74. Osinga T, Frommherz U, Steinfeld A, Wieckert C. Experimental Investigation of the Solar Carbothermic Reduction of ZnO Using a Two-Cavity Solar Reactor. Journal of solar energy engineering. 2004;126(1):633-7.
75. Epstein M, Olalde G, Santén S, Steinfeld A, Wieckert C. Towards the Industrial Solar Carbothermal Production of Zinc. Journal of Solar Energy Engineering. 2008;130(1):014505.
76. Wieckert C, Frommherz U, Kräupl S, Guillot E, Olalde G, Epstein M, et al. A 300 Kw Solar Chemical Pilot Plant for the Carbothermic Production of Zinc. Journal of Solar Energy Engineering. 2007;129(2):190-6.
77. Piatkowski N, Steinfeld A. Solar-Driven Coal Gasification in a Thermally Irradiated Packed-Bed Reactor. Energy & Fuels. 2008;22(3):2043-52.
78. Kromer M, Roth K, Takata R, Chin P. Support for Cost Analyses on Solar-Driven High Temperature Thermochemical Water-Splitting Cycles. Lexington, MA: TIAX, LLC2011 February 22, 2011. Report No.: DE-DT0000951.

6

CHAPTER

PHOTOVOLTAIC WATER PUMPING SYSTEM IN NIGER

Madougou Saïdou[1], Kaka Mohamadou[1] and Sissoko Gregoire[2]

[1] *Université Abdou Moumouni de Niamey, Niger*

[2] *Université Cheikh Anta Diop de Dakar, Senegal*

1. INTRODUCTION

Solar energy such as photovoltaic is the most promising energy of the non-conventional energy sources which is capable to satisfy the energy needs of the isolated rural areas. It fits perfectly to the decentralization of power generation for the small communities widely dispersed as evidenced the solar pumps whose operation is nowadays proved flawless (World Resources Institute, 1992).

The use of photovoltaic energy to pump water is particularly well suited in the Sahel. This source of energy is free and abundant, but also provides autonomy for many isolated villages of rural areas. The water pumped is stored in the thanks until its use (in the night or during the cloudy days). Locally, there are many companies that manufacture these tanks. These towers do not require special maintenance and are easy to be repair. This chapter presents a study of photovoltaic water pumping in Niger rural areas. We present first the benefits of photovoltaic water pumping; we describe in the second the photovoltaic water pumping system, before present the solar radiation at Niger. Then the sizing of a photovoltaic water pumping system is described and finally, a case study of photovoltaic water pumping is presented before concluding.

2. BENEFITS OF PHOTOVOLTAIC WATER PUMPING SYSTEM

In African countries, many governments are still struggling to meet the basic needs of the population due to lack of availability of electric power system. In rural areas, these needs are summarized in the drinking water, electrification of health centers and irrigation. For lack of resources, these people are resolved to supply water from wells craft of 10 to 100 m and less than 2 m of diameter. Water is the source of life. Its rarity is one of the tragedies of the Sahel. In this area, people have regularly serious problems of drinking water. The water which involves in agricultural and domestic consumption requires dewatering technologies adapted to local conditions. Yet, the solar energy potential is very abundant. The geographical situation of Niger fosters the development and growth of the use of solar energy. Indeed, given the importance of the solar radiation and the duration of sunshine, which excesses eight hours (8 hours) per day throughout the year, some electricity needs can be met by solar energy. This energy source may be beneficial to the most remote areas, especially in water pumping. The use of solar energy can make a meaningful and lasting solution to the drainage in rural areas. The use of photovoltaic solar energy for water pumping is well suited to these regions due to the existence of a potential groundwater quite important especially in desert areas, and a large solar energy potential. Also, this solar energy has the advantage of being clean compared to conventional energy sources which present constraints of remoteness from the mains, fuel transport and periodic maintenance for diesel engines (Jimmy et al 1998, Barlow, 1993). Photovoltaic solar energy may well contribute to meet the energy needs of these populations. Photovoltaic generators are very suitable to relieve the lack of availability of grid. Indeed, with an installation of 1 kW maximum power, we can pump water to heights of about 80 m, or meet irrigation needs of a village (Commission of The European Community, 1985). In addition, these solar generators can be used concomitantly using batteries to power television sets, or for lighting when the pumping system is not loaded (Billerey, 1984).

3. PHOTOVOLTAIC PUMPING SYSTEM

3.1. Description

Photovoltaic pumping system is a standard pump equipped with an electric motor, provided in electrical energy by photovoltaic panels installed on the site (Handbook on Solar Water Pumping, 1984; Fraenkel and al., 1986).

This pump is intended to pump water from the basement to make it accessible to users (Fig. 1).

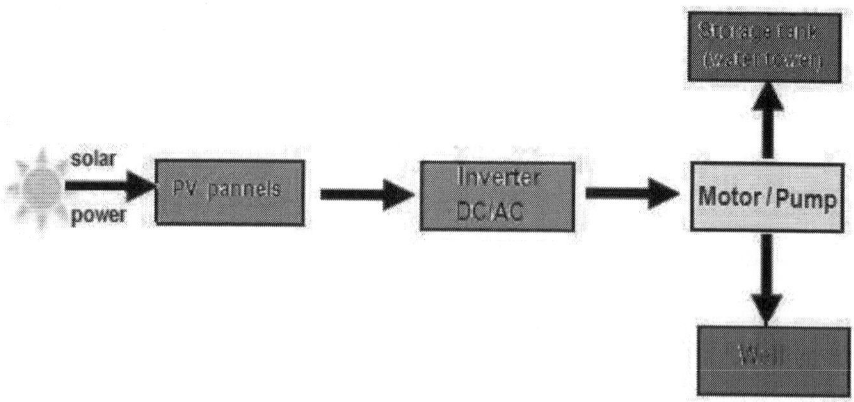

FIGURE 1. Principe of photovoltaic pumping system.

Nowadays, two types of photovoltaic water pumping systems are used: the photovoltaic water pumping with batteries and without batteries. In Niger, it is often used photovoltaic water pumping without batteries, commonly known as "pumping over the sun". Pumping over the sun is simpler and less expensive than with battery system. Instead of batteries, they use a tank to store water until it is used (Fig.2). Hydraulic storage allows overcoming electrical energy storage thus avoiding the use of batteries which have a limited life (6 years compared with 20 to 30 years of photovoltaic panels) and are polluting. However, the method without batteries has some drawbacks and its main fault is to have a flow of water which depends on the average time of the sunlight (A. Hadj, 1999).

There are two types of pumps to draw water surface: Positive Displacement *Pumps* (volumetric pumps) and centrifugal pumps. Besides to the type of pump, there are two other characteristics at the pumps according to the physical location of the pump in relation to the pumped water: the suction system and stuffer one. They discharge pumps are submerged in water. Their motor is immersed in water with the pump and the discharge pipe placed after the pump can lift water to tens of meters to the storage tank depending to the engine's power. Afterward, the system is connected to a distribution network that delivers water to users.

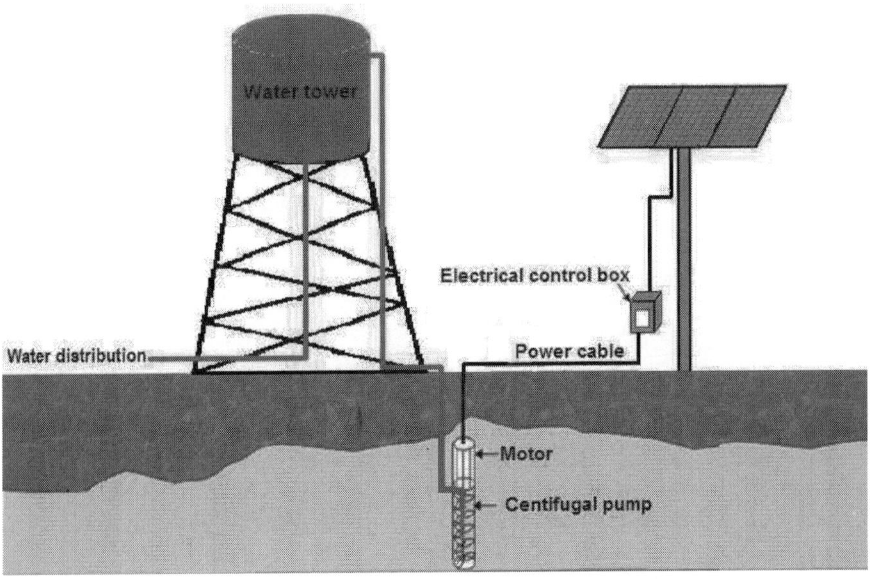

FIGURE 2. Photovoltaic water pumping without batteries with a tank to store water.

3.2. PHOTOVOLTAIC SYSTEM

To generate the necessary energy to the motor of the pump, solar photovoltaic panels are placed for converting solar energy into electrical energy (Fig.3). As the panels generate a direct current (DC), it is often used DC/AC converter to convert the direct current produced by the solar panels into alternative current (AC) if the motor of the pump is AC. On the other side, if the motor is DC, the device does not need a DC/AC converter. The energy produced by the panels can be used directly or stored. In the case of an application for water pumping, it is more interesting to use the energy to raise the water in a castle that serves as hydraulic energy storage.

To prevent a dysfunction of the pump when it is live on photovoltaic, due to under sizing or over sizing the PV generator, an inverter is used to ensure the proper operation of the PV/pump system.

FIGURE 3. Students determining the current-voltage characteristic I = f(V) of solar cell "over the sun" at the University Abdou Moumouni of Niamey (Niger).

4. SOLAR RADIATION AT NIGER

To design a photovoltaic water pumping system, we will need to quantify the available solar energy. Therefore, it is very important to know the solar radiation of the locality. Solar radiation (kW/m^2) is the energy from the sun that reaches the earth. The earth receives a nearly constant of solar radiation at its outer atmosphere. The intensity of solar radiation varies with geographic location (Fig.4).

It also varies with the season and time of the day. In Sahel, the solar radiation at the earth's surface is very important (Fig.5).

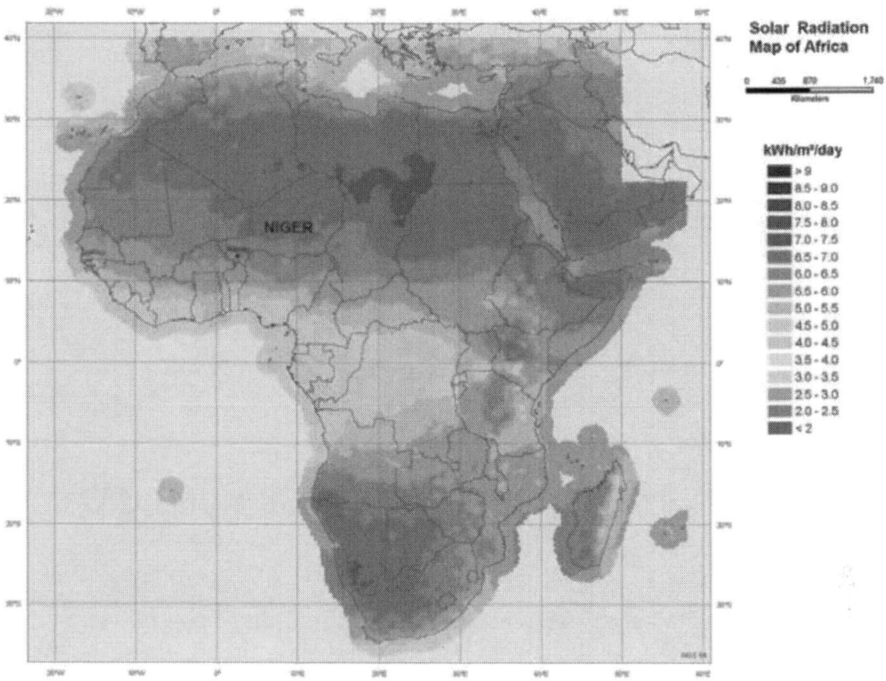

FIGURE 4. Africa solar radiation map (Source: UNEP, NREL and the Global Environment Facility).

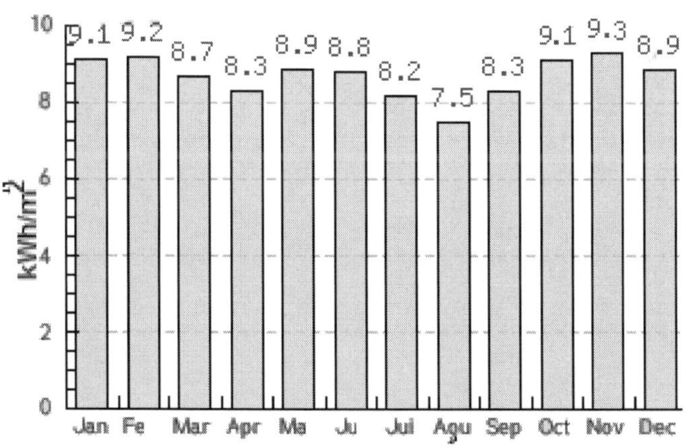

FIGURE 5. Daily average radiation of sunlight per month at Niger ($kWh/m^2.day$).

The most productive hours of sunlight are from 9:00 a.m. to 5:00 p.m. Table 1 gives the daily average time (Hours/day) of the sunlight at Niger (Keita town: Latitude 14.75°N and Longitude 5.76°E).

TABLE 1. Monthly average time (Hours/days) of the sunlight at Niger (Keita town: Latitude: 14.75°N and Longitude: 5.76°E)

Month	Jan.	Feb.	Mar.	Ap.	Ma	June	July	**Aug.**	Sept.	Oct.	Nov.	Dec.
Average Time (Hours/ day)	9	9.3	8.7	8.4	8.8	8.6	8.2	**7.5**	8.1	9.2	9.5	8.9

5. SIZING A PHOTOVOLTAIC WATER PUMPING SYSTEM

Photovoltaic water pump sizing is the determination of the power of the solar generator that will provide the desired amount of water (Alonzo, 2003).

The photovoltaic water pump sizing consists of:

- Assessment of daily water needs of the population to know the rate flow required;
- calculation of hydropower helpful;
- determining of the available solar energy;
- determining of the inclination of the photovoltaic generator which can be placed;
- determination of the month sizing (the month in which the ratio between solar radiation and hydropower is minimum);
- sizing of the PV generator (determination of the required electrical energy);

5.1. Assessment of Daily Water Needs of the Population

Determining the water needs of the consumption of a population depends mainly of its lifestyle, the environment and climatic conditions of each region. Drinking, cooking, washing and bathing are the main uses of water for human needs. Animals also need water for their survival. The water use is also essential in the field of agriculture.

Depending on the nature of the users (humans, animals) or the use, the amount of water required for each user or usage are:

Population (Humans)	– 5 liters /day, for survival;	
	– 10 liters /day, for the minimum acceptable;	
	– 30 liters /day, the normal living conditions in Africa;	
Animals	– Cattle	40 liters /day;
	– Sheep, Goat	4 liters /day;
	– Horse	40 liters /day
	– Ass	20 liters /day
	– Camel	20 liters /day (reserve for 8 days)
Irrigation (agriculture)	– Crops at the village	60 m^3 /day / hectare;
	– Rice	100 m^3 /day / hectare;
	– Cereals	45 m^3 /day / hectare;
	– Sugar cane	65 m^3 /day / hectare;
	– Cotton	55 m^3 /day / hectare.

There are three standards for the calculation of water requirements:

1. The standard for the minimum amount for survival.
2. The current target of funding agencies: 20 liters/day/person that does not include livestock and gardening.
3. The minimum amount necessary for economic development: 50 liters/day/person. It includes:

- 20 liters/day/person, for personal needs;
- 20 liters / person / day, 0.5 unit of cattle per person;
- 10 liters / day / person, 2 m^2 for vegetable crops.
-

In applying the standards 2 and 3, we obtain the following water requirements (Table 2) depending on the size of the village (population).

TABLE 2. Water requirements depending on the size of the village (population).

Population of village (persons)	Standard 2: (20 liters/day/person) (m^3)	Standard 3: (50 liters/day/person) (m^3)
250	5	12.5
500	**10**	**25**
750	15	37.5
1000	20	50
1500	20	75
2000	40	100

5.2. Determination of Hydropower Helpful

The average daily load i.e. hydropower helpful (kWh/day) required is expressed by:

$$E_H = \frac{g^*\rho_a{}^*Q_a{}^*TH}{\eta_P{}^*3600} = \frac{C_H{}^*Q_a{}^*TH}{\eta_P} \quad (1)$$

Where, g is acceleration of gravity (9.81 $m.s^{-2}$);

ρ_a is water density (1000 kg/m^3);

Q_a is daily water needs (m^3/day);

TH is the total head (m);

η_P is pump system efficiency

The tank capacity is determined by the daily water needs and the autonomy of the system.

5.3. Determining of the Available Solar Energy

The method used is based on the determination of daily mean values of solar radiation available and hydropower necessary.

5.4. Determining of the Inclination of the Photovoltaic Generator which can be Placed

The inclination β to the horizontal plane of the photovoltaic panels (PV) must be to maximize the relationship between solar radiation and hydropower necessary. We have chosen 15°N, the latitude of the locality (Keita town: Latitude 14.75°N).

5.5. Determination of the Month Sizing

The sizing month will be the worst month, i.e., the month that the ratio between solar radiance and hydraulic energy required is minimal. In our case it is the month of August is the month of sizing. In august the average time is 6.5 hours/day.

5.6. Sizing of the PV Generator

As the system works all year round, the field is tilted at an angle equal to the latitude of 15 ° N. It was in August that the average number of hours of sunshine is the lowest maximum: 6.5 hours/day. Assuming a 25% loss due to the temperature and dust, the required electrical energy is given by:

$$W_{PV} = \frac{E_H}{Radiance*(1 - loss)} \tag{2}$$

6. CASE STUDY

A study on the photovoltaic water pumping system in a village at 30 km of Keita (Niger) to meet the water needs of the five hundred (500) persons gave the following summarized results.

FIGURE 6. Photovoltaic water pumping station with a volumetric pump at Niger.

a. Assessment of daily the water needs: Using the Standard 3: (50 liters/day/person), the water needs rises to 25 m^3/day.

b. The average daily load i.e. hydropower helpful (kWh/day) required is given by this expression :

$$E_H = \frac{g^* \rho_a {}^* Q_a {}^* TH}{\eta_P {}^* 3\,600} = \frac{C_H \cdot Q_a {}^* TH}{\eta_P}$$

With g = 9.81 m.s^{-2}

ρ_a = 1000 kg/m^3

Q_a = 25 m^3/day

TH = 52 m

η_P = 50 %

It provides: EH=7 085

a. The available solar energy:

b. Daily average radiation of sunlight varies from 7.5 to 9.3 kWh/m^2/day.

To make sure to do a good sizing, we choose the minimum value of average radiance: 7.5 kWh/m2/day.

The inclination to the horizontal plane of the photovoltaic panels is: β=15°.

a. The sizing month is: August, 6.5 hours/day.

b. Sizing of the PV generator

Assuming a 25% loss due to the temperature and dust, the required electrical energy is given by this expression:

$$W_{PV} = \frac{E_H}{Radiance^*(1 - loss)}$$

$$W_{PV} = 1260 \text{ Wc}$$

The operating point of our photovoltaic field is set at 120 volts due to the characteristics of the inverter. The photovoltaic field will be composed of 10 multiple modules in series. Generator power is 1260 Wc, the rate current is 10.50 A. With photovoltaic panels which have 3.5 A, we will have 3 modules in parallel.

The table 3 shows the summary of case study.

TABLE 3. Summary of case study

Month of sizing	**Radiance (kWh /m²)**	**Time of sunlight (H)**	**Loss (%)**	EHEH(**kWh)**	WPVWP V**(Wc)**	**Voltage (V)**	**Current (A)**	**Configuration of panels**
August	7.5	7.5	20	7 085	1260	120	3.5	10 X 3

7. CONCLUSION

The use of solar energy in Niger, particularly photovoltaic energy, for water pumping is well suited in this arid and semi-arid area due to the existence in this region of an underground water potential, and a large solar energy potential more than 6 kWh/m^2.

Photovoltaic generators are coupled directly to the pump with a DC/AC converter. Storing water in the tanks avoids additional costs accumulator used to store electrical energy. The case study clearly shows the advantage of photovoltaic pumping system compared to conventional energy one which has many constraints of distance to the power grid, of transportation of fuel, and of periodic maintenance of the engines. The cost of one cubic meter (1 m^3) of water pumped by the PV system is more advantageous than others systems. This pumping system constitutes a solution for the water supply of these sparsely populated, remote and isolated areas (Thomas, 1987). With the falling prices of solar panels, this source of energy must be popularized and integrated in the development strategy of these countries.

REFERENCES

1. A Hadj, F Arab, K Chenlo, Mukadam and J.L. Balenzategui, (1999Performance of PV Water Pumping Systems". Renewable Energy, 18N°2, 191204
2. R Barlow, B Mcnelis, Et Derrick, A, (1993Solar Pumping: An Introduction and Update on the Technology, Performance, Costs, and Economics, World Bank Technical Paper 168Intermediate Technology Publications, London, UK.
3. J Billerey, 1984Le pompage photovoltaïque, GRET, Paris, France.
4. Commission of The European Community1985Renewable Sources of Energy and Village Water Supply in Developing Countries. Directorate General for Development, EEC.
5. C Alonzo, 2003Contribution à l'Optimisation, la Gestion et le Traitement de l'Energie ». Thèse de Doctorat, Toulouse.
6. P Fraenkel, 1986Water Pumping Devices, I.T Power, London, UK.
7. Handbook on Solar Water Pumping (1984Intermediate Technology Power & Partners, eading, 124 p.

8. R Jimmy, D Thomas, E Schiller, and B. S Sy, 1998Le pompage photovoltaïque : manuel de cours à l'intention des ingénieurs et des techniciens. Éditions Multi Mondes. 254 p. 2-89481-006-7
9. M. G Thomas, 1987Water Pumping- The Solar Alternative". Photovoltaic Design, Assistance Center. Sandia National Laboratories. NM 87185, Albuquerque, 58 p.
10. UNEPNREL and the Global Environment Facility (2012www.unep.org/dgef
11. World Resources Institute1992Ressources mondiales 1992-1993 : un guide de l'environnement global, PNUD, Éditions Sciences et Culture, Montréal, Canada.

7

CHAPTER

COST AND EMISSIONS IMPLICATIONS OF COUPLING WIND AND SOLAR POWER

Seth Blumsack, Kelsey Richardson

John and Willie Leone Family Department of Energy and Mineral Engineering, Pennsylvania State University, University Park, USA.

ABSTRACT

We assess the implications on long-run average energy production costs and emissions of CO_2 and some criteria pollutants from coupling wind, solar and natural gas generation sources. We utilize five-minute meteorological data from a US location that has been estimated to have both high-quality wind and solar resources, to simulate production of a coupled generation system that produces a constant amount of electric energy. The natural gas turbine is utilized to provide fill-in energy for the coupled wind/solar system, and is compared to a base case where the gas turbine produces a constant power output. We assess the impacts on variability of coupled wind and solar over multiple time scales, and compare this variability with regional demand in a nearby load center, and find that coupling wind and solar does decrease variability of output. The cost analysis found that wind energy with gas back-up has a lower levelized cost of energy than using gas energy alone, resulting in production savings. Adding solar energy to the coupled system increases levelized cost of energy production; this cost is not made up by any reductions in emissions costs.

Keywords: Wind Energy; Solar Energy; Air Emissions

1. INTRODUCTION

The integration of renewable electric generation resources, particularly solar and wind, into existing energy production and delivery systems, is considered to be a major technical and economic challenge. Variability of both wind and solar resources introduces additional stochastic components to the operational objective of balancing supply and demand across power control areas on a real-time basis; and presents planning challenges to system operators and generating companies that must make investment portfolio decisions to accommodate increased renewable energy penetration. Integration costs are expected to rise along with penetration levels, with recent reports suggesting costs of \$2/MWh to \$9/MWh [1] once wind penetration reaches 20% [2].

A number of different operational strategies are possible to achieve large-scale penetration of wind and solar energy. One such strategy would expand the use of frequency regulation and balancing energy services to compensate for unforecasted deviations in wind and solar power output. These so-called ancillary services are currently utilized in large-scale electricity systems to balance unexpected deviations in the supply-demand balance caused by volatile electricity demand or unplanned outages at generation or transmission assets. Such ancillary services can be provided on the supply side or the demand side. While system operators are experienced in procuring and dispatching ancillary services to handle fluctuations in the supply-demand balance, it is less clear whether existing ancillary services, and the various types of markets through which they are procured, are adequate to support the large-scale integration of wind and solar resources while maintaining high reliability and reasonable costs [3].

Wind and solar resources could also self-provide balancing energy by coupling their output with a controllable resource to provide a firm energy schedule to the system operator. The controllable resource essentially provides some amount of "fill-in" power to compensate for the variable resource. This coupling could be contractual, in which the wind or solar resource signs a physical energy purchase contract with a complementary resource; or the coupling could occur through horizontal integration, where a single generation operator manages the output of the variable and controllable resource to provide the firm energy schedule. The provision of fill-in power for wind turbines has been studied for hydroelectric dams [3], compressed air energy storage [4], and hybrid fill-in systems consisting of natural gas turbines that compensate for "slow" fluctuations and battery energy storage that compensate for "fast" energy storage [5].

The increased use of ancillary services versus the coupling of variable and controllable resources is an unsettled topic in the electric utility industry and in the relevant research literature. We choose to assess the direct coupling of variable and controllable resources; in particular we assess how coupling multiple variableresources affects the utilization of the controllable resource. Specifically, we compare the costs and emissions of two coupled energy systems with a base-case system featureing a natural gas combustion turbine that produces a constant amount of electric energy. For comparison, we couple the natural gas turbine with an equivalently-sized wind energy installation; as well as coupling the gas turbine with a wind energy installation and a solar photovoltaic installation. Differences in the diurnal cycles of wind and solar energy [3] suggest that coupling these two variable sources may reduce the utilization of the controllable (and polluting) energy resource, with concomitant benefits in terms of air emissions.

The remainder of the paper discusses our data sources, modeling strategy and results. Section 2 outlines the data we have obtained for a high-quality wind and solar site in western Oklahoma, and describes how we estimate wind and solar production based on meteorological data. Technology and emissions costs are also discussed in Section 2. Section 3 contains the results of our modeling exercise, and Section 4 offers some discussion and conclusions.

2. DATA AND MODELING

Our case study is based on simulation of a coupled system of wind, solar and natural gas installed in Weatherford, in western Oklahoma. Weatherford is considered an advantageous location for such a coupled system since it is located in a high-quality wind zone (Class IV wind site [6]) and has relatively high-quality photovoltaic potential (based on analysis in [7]). We collect five-minute meteorological data for the Weatherford location from the Oklahoma Mesonet, a collection of over one hundred measurement stations across Oklahoma. Data available through the Mesonet includes averages of five-minute wind speed (m/s), wind direction, average solar radiation (W/m^2), and other weather variables. Data are recorded at a height of 10 m. We use 2002 as our study year, since that year is considered to be a historically average wind and solar irradiance year for Oklahoma.

2.1. Wind Power Modeling

Five-minute wind speed data at 10 m are converted to 80 m hub height data using the power-law equation:

$$V_{2,t} = V_{1,t}\left(H_2/H_1\right)^{1/7}, \tag{1}$$

where $V_{1,t}$ and $V_{2,t}$ are the recorded and estimated wind speeds (m/s) at time interval t, at heights H_1 and H_2 (so in our case, H_1 = 10 m and H_2 = 80 m; $V_{1,t}$ represents the time series of wind speeds that we obtained from the Oklahoma Mesonet; and $V_{2,t}$ represents the estimated time series of wind speeds). As a robustness check, we generated estimated wind speeds using the roughness equation:

$$V_{2,t} = V_{1,t}\left\{\ln\left(H_2/l\right)/\ln\left(H_1/l\right)\right\}, \tag{2}$$

where l is the roughness length. We do not have explicit data on the roughness length for our location in Weatherford, although estimated wind speeds using a roughness length of 0.03 (corresponding to open and largely flat agricultural area) did not significantly differ from those estimated using the power-law equation in (1).

Finally, we converted our estimated wind speed data at a hub height of 80 m to wind power data using a power curve for a GE 1.5SL turbine; the power curve was obtained from Idaho National Laboratory data [8], and is shown in **Figure 1**. We note here that since our original Mesonet wind speed data are five-minute averages, our power calculations should be taken to represent average power within any given five minute interval.

Based on our simulated wind power production data, we calculated average energy (kWh) produced by a single turbine during each five minute interval for the Weatherford location. We calculated an annual capacity factor of 0.4 for the Weatherford wind site, based on five-minute data. Variation of estimated wind power production over several cycle lengths is shown in **Figure 2**, in terms of standard deviation of output and volatility (percentage standard deviation); the results are broadly consistent with previous analyses of the power spectral density of wind turbine outputs [8,9].

To examine the correlation between estimated wind energy output at Weatherford and electricity demand in Oklahoma, we obtained hourly electricity demand data for Oklahoma Gas and Electric (OGE) and Oklahoma Municipal Power Authority from FERC Form 714 [10]. While OGE and OMPA are not the only electric utilities in Oklahoma, data from other

Oklahoma utilities was not available for our analysis. A complete data set of hourly electricity demand in Oklahoma is not vital for our analysis, since we are primarily concerned with the relative variation in demand compared to wind and solar energy production.

We normalized both hourly wind production and hourly electricity demand for OGE and OMPA on a seasonal basis; this normalized data is shown in **Figure 3**. The annual correlation between wind production and electricity demand, based on analysis of the five-minute data, is 0.061. We note that unlike analyses of wind energy production and electricity consumption in other areas of the US [3,11], there is a small but positive degree of correlation between our estimates of wind energy production in Weatherford and electricity consumption in portions of Oklahoma. We also observe more diurnal variability, on average, during the summer months than during other times of the year.

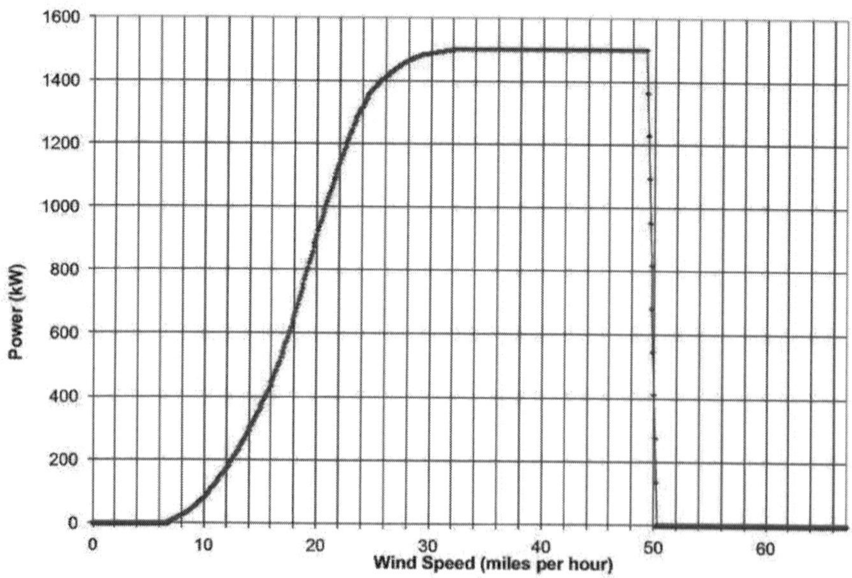

Figure 1. Wind power curve for the GE 1.5SL turbine.

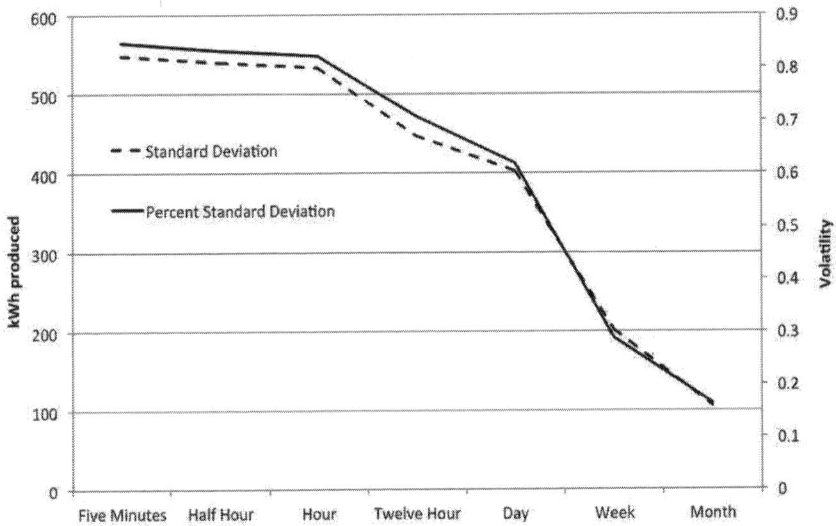

Figure 2. Variation in energy production (kWh, left axis) and percentage variation in production (volatility, right axis) for the simulated wind site in Weatherford.

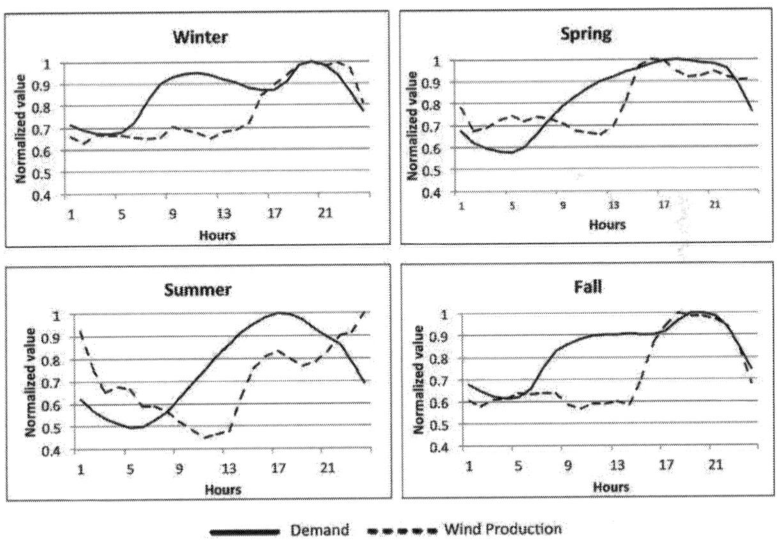

Figure 3. Normalized average wind production at Weatherford and electricity demand in the OGE and OMPA service territories.

2.2. Solar Production Modeling

Five minute solar radiation data (W/m^2) are utilized along with TRNSYS, an energy simulation program, to simulate photovoltaic energy production in Weatherford[1]. We assume a flat-plate fixed-axis collector with tracking, and a conversion efficiency of 15%. A variety of factors affect the production of energy from photovoltaic modules apart from the level of solar irradiance including dust and haze. We utilized detailed meteorological data from the ASOS station at Stafford Airport, a short distance east of our measurement station for Weatherford to incorporate the impacts of dust on solar energy production. The simulated photovoltaic installation has an estimated capacity factor of 0.17.

We compared normalized photovoltaic energy production with normalized electricity demand and wind energy production by season, as shown in **Figure 4**. We observe the typical seasonal variations in photovoltaic production, with an extended period of peak photovoltaic production in the summer compared to other seasons. Graphically, we do not observe anti-correlation between wind and solar production of a particularly large magnitude; the seasonal correlations between wind and solar production for our case study are 0.27 in the wintertime, 0.2 in the spring, 0.21 in the summer and 0.21 in the fall.

2.3. Natural Gas Production Modeling

The natural gas turbine that we use in our analysis is a GE 10-1 combustion turbine with a nameplate capacity of 10 MW and a combustion efficiency of 31.4%[2] at the time of this writing, the price of natural gas in Oklahoma was $6.96 per million BTU [12], implying a marginal energy cost of production of $75.63 per MWh. A heat rate curve for this specific turbine was not available, so we use the quadratic emissions model described in [13] to estimate emissions of NO_x and CO_2 as a function of the magnitude and speed with which production at the gas turbine is ramped up and down in response to fluctuations in solar and energy resource availability. While the method in [13] has been criticized [14], it seems appropriate given the scale of our problem (i.e., powerplant level operations).

2.4. Long Run Average Cost

The Levelized Cost of Electricity (LCOE, units of $/MWh) represents the long-run average cost of a power plant. We use the LCOE metric to compare the costs of a natural gas turbine with the costs of two coupled generation installations. The first system couples output from a wind installation with the gas turbine, while the second couples output from wind and photovoltaic installations with the gas turbine. The LCOE measures the average price required to break even over the long run—i.e., to recover the present discounted value of all capital and operating costs. The LCOE is defined as [15]:

$$\mathrm{LCOE} = \left\{ (r \times FC) / 1 - e^{-rt} \right\} / cf + VC, \tag{3}$$

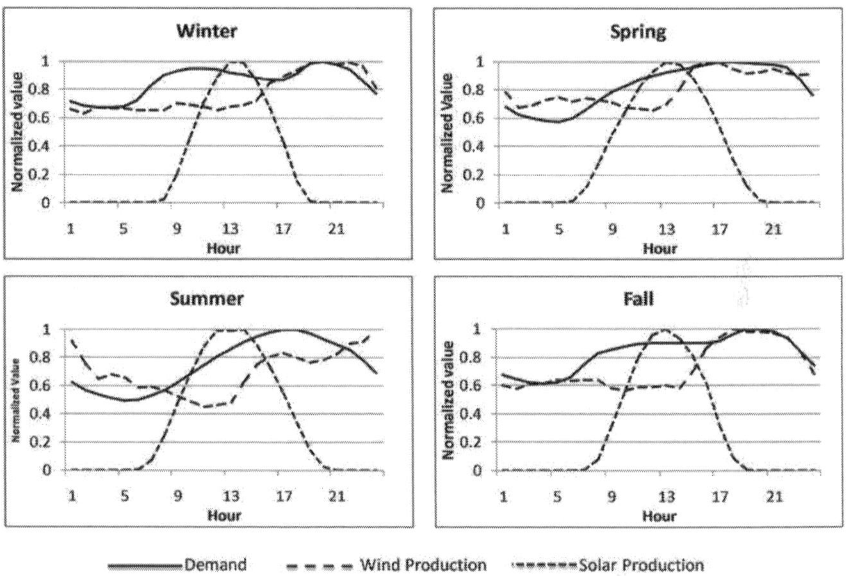

Figure 4. Normalized average wind and solar energy production at Weatherford and electricity demand in the OGE and OMPA service territories.

where r is the annual discount rate; FC represents overnight fixed costs, T is the relevant time horizon for the project (assumed to be 20 years for all technologies), cf is the project's capacity factor, and VC is the variable cost of producing a unit of energy. We assume that the overnight fixed costs are \$500/kW for the natural gas plant [16], \$2100/kW for the wind turbine [6] and \$5000/kW for the photovoltaic installation [16].

3. RESULTS

We simulate the operation of three systems producing a constant amount of power, and compare the cost and emissions implications (CO_2 and NO_x) of each. The three systems we consider in our analysis are a standalone gas combustion turbine; a coupled system of wind and natural gas; and a coupled system of wind, solar photovoltaic and natural gas. In the latter two scenarios, the gas turbine is utilized for providing fill-in power as described in Section 1. The cost and emissions implications are compared on a per-unit-energy basis, so we do not need to specify an explicit size for the system (although our capital cost figures in Section 2.4 implicitly assume utility-scale installations).

Given the capital and operating costs of the gas and wind turbines, and the high capacity factor of the modeled wind installation in Weatherford, the LCOE for the coupled wind/gas system is lower than for the gas system alone ($86.59/MWh for the coupled system versus $95.60/MWh for the natural gas turbine alone). The LCOE of the coupled system is naturally sensitive to fluctuations in the fuel price for the gas turbine and the availability of the wind turbine; if the price of natural gas were to drop below $6/mmBTU then the gas turbine alone would be more economical. Because of the high capital costs of solar photovoltaic, and the low capacity factor, we find that coupling photovoltaics with the wind/gas system increases the LCOE to $223.24/MWh. Defining the integration cost for solar photovoltaics as the increase in (levelized) energy costs for each fiveminute interval for the wind/solar/gas system versus the wind/gas system, we see that average integration costs for including photovoltaics in a portfolio of wind and fossil energy vary seasonally and range from between $57/MWh to nearly $120/MWh (**Figure 5**).

Moreover, the emissions of NO_x following the integration of photovoltaics into the wind/gas system increase for the scenario that we simulated, although CO_2 emissions from the coupled system decline. **Figure 6**(a) shows that integrating wind and solar with the natural gas turbine reduces CO_2 emissions by 500 to 1000 tonnes per month, assuming a system with 10 MW of renewables and 10 MW of natural gas turbine capacity. The incremental avoided emissions from coupling photovoltaics with the wind/gas system are relatively small, at 25 tonnes per month. **Figure 6**(b) shows NO_x emissions from the three systems; coupling wind and solar with the natural gas turbine increases NO_x emissions by a factor of between three and five, depending on the time of year.

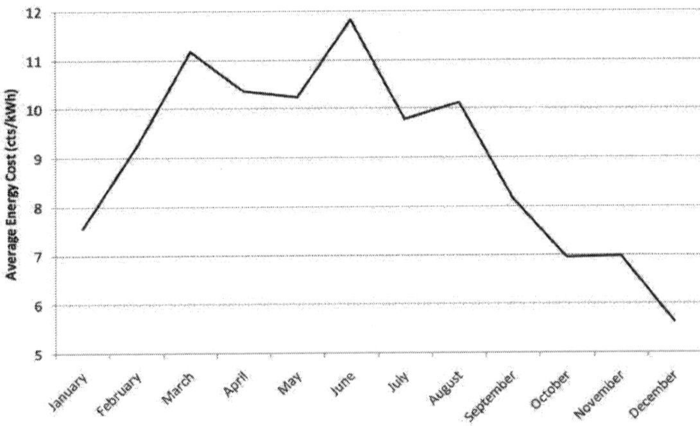

Figure 5. Average energy cost of integrating photovoltaics into the wind/gas portfolio.

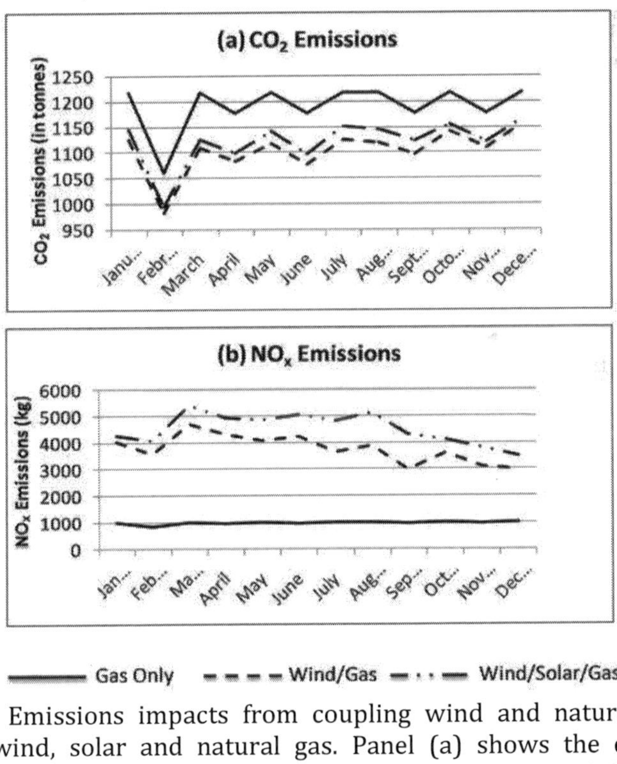

Figure 6. Emissions impacts from coupling wind and natural gas; and coupling wind, solar and natural gas. Panel (a) shows the decrease in CO_2emissions from coupling renewables and fossil fuels, while panel (b) shows the increase in NO_x emissions.

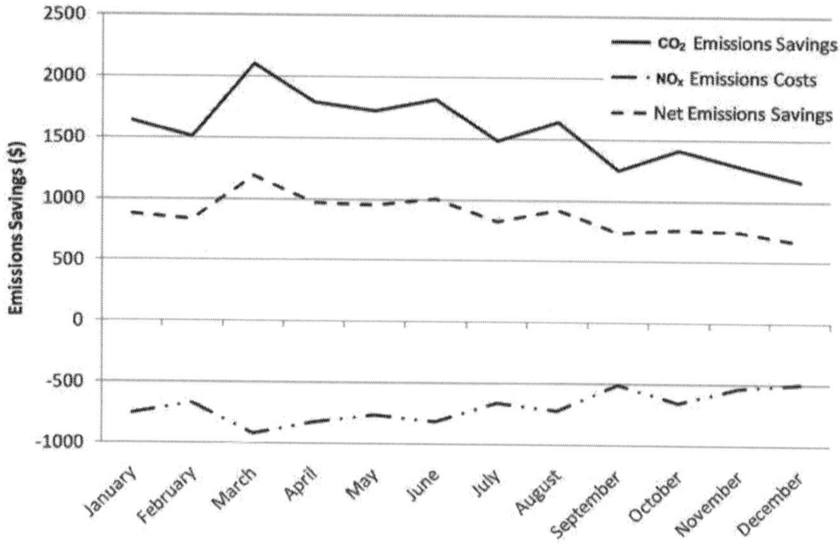

Figure 7. Monetary savings, including emissions costs, from coupling wind and natural gas. We do not find any monetary savings from adding photovoltaics to the wind/gas capacity portfolio.

4. DISCUSSION AND CONCLUSIONS

Our results on the emissions implications of coupling renewables and fossil fuels are consistent with previous research [13]. Coupling wind energy with a natural gas turbine can potentially reduce long-run average production costs, although incrementally adding photovoltaics to the portfolio increases costs. We find that the coupled wind/gas system has higher NO_x emissions than simply running a natural gas turbine at a constant level of output, but lower CO_2 emissions. Adding photovoltaics reduces the CO_2 emissions profile of the system slightly while increasing the NO_x profile.

In a scenario where emissions are priced, the reduction in CO_2 cost may outweigh the increase in NO_x costs for the wind/gas coupled system, yielding net savings (compared to running the gas-only system) depending on the relative emissions prices. Using prevailing CO_2 prices from the European Market of $22.60 per tonne, and a NO_x price of $0.25/kg, we find that pricing emissions leads to additional net savings for the wind/gas coupled system (**Figure 7**). While there would be additional net emissions cost savings from adding photovoltaics to the coupled system, the savings are not sufficient to offset the higher long-run average production cost.

The incentives for owners of renewable energy projects to couple operations with controllable energy sources are still a matter of debate. In some territories that have adopted wholesale restructuring, such coupling would allow the owners to receive capacity payments. In territories where bilateral market activity is the dominant form of trade, coupled renewable and fossil projects may be able to sign "firm" energy contracts (subject to availability of transmission), which typically sell at a premium.

5. ACKNOWLEDGEMENTS

The authors would like to acknowledge support from the Wilson Endowment in the College of Earth and Mineral Sciences at Penn State for the purchase of Oklahoma Mesonet data; staff at the Oklahoma Mesonet project for assistance with the data; and Jeffrey Brownson and Rebecca Hott for advice on modeling solar energy production.

REFERENCES

1. NREL, "Wind Technologies Market Report," 2010. www1.eere.energy.gov/wind/pdfs/51783.pdf
2. US DOE, "20% Wind Energy by 2030: Transmission and Integration into the US Electric System," 2008. http://www.20percentwind.org/Final_DOE_Executive_Summary.pdf
3. A. Fernandez, S. Blumsack and P. Reed, "Evaluating Wind-Following and Ecosystem Services for Hydroelectric Dams," Center for Research in Regulated Industries Eastern Conference, Skytop, May 2011.
4. E. Fertig and J. Apt, "Economics of Compressed Air Energy Storage to Integrate Wind Power: A Case Study in ERCOT," Energy Policy, Vol. 39, No. 5, 2011. pp. 2330- 2342. doi:10.1016/j.enpol.2011.01.049
5. E. Hittinger, J. Whitacre and J. Apt, "Compensating for Wind Variability Using Co-Located Natural Gas Generation and Energy Storage," Energy Systems, Vol. 1, No. 4, 2010, pp. 417-439. doi:10.1007/s12667-010-0017-2
6. NREL, "Wind Energy Resource Atlas of the United States," Renewable Resource Data Center (RReDC) Home Page, 2002. http://rredc.nrel.gov/wind/pubs/atlas/appendix_A.html

7. University of Oklahoma, Environmental Verification and Analysis Center. http://www.ocgi.okstate.edu/owpi/Oksolarmap.gif
8. Idaho National Laboratory, "Wind Turbine Power Curve Data," 2012. http://www.inl.gov/wind/software/
9. J. Apt, "The Spectrum of Power from Wind Turbines," Journal of Power Sources, Vol. 18, No. 2, 2007, pp. 369- 374. doi: 10.1016/j.jpowsour.2007.02.077
10. FERC, "Form 714-Pre-Electronic Filing Data: 1993- 2004," Federal Energy Regulatory Commission, 2011. http://www.ferc.gov/docs-filing/forms/form-714/data.asp
11. Carnegie Mellon Electricity Industry Center, "The Smart Grid: Sorting the Reality from the Hype," 2009. http://www .cmu.edu/electricity
12. Price data from US Energy Information Administration, 2012. http://www.eia.gov/dnav/ng/ng_pri_sum_dcu_nus_m.htm
13. W. Katzenstein and J. Apt, "Air Emissions Due to Wind and Solar Power," Environmental Science & Technology, Vol. 43, No. 2, 2009, pp. 253-258. doi:10.1021/es801437t
14. A. Mills, R. Wiser, M. Milligan and M. O'Malley, "Comment on 'Air Emissions Due to Wind and Solar Power'," Environmental Science and Technology, Vol. 43, No. 15, 2009, pp. 6106-6107. doi:10.1021/es900831b
15. S. Stoft, "Power System Economics: Designing Markets for Electricity," Piscataway, IEEE, 2002.
16. Energy Information Administration, "Updated Capital Cost Estimates for Electricity Generation Plants," 2011. http://www. eia.gov/oiaf/beck_plantcosts/pdf/updatedplantcosts.pdf

8 CHAPTER

SUSTAINABILITY IN SOLAR THERMAL POWER PLANTS

Rafael Almanza[1] and Iván Martínez[1]

[1] Universidad Nacional Autónoma de México / Universidad Autónoma del Estado de México, México

1. INTRODUCTION

In the last two decades, there has been growing interest in developing indicators to measure sustainability, which is currently seen as a delicate balance between the economic, environmental and social health of a community, a nation or even our planet. The current measure of sustainability tends to be an amalgam of economic, social and environmental aspects. Economic indicators have been used to measure the state of regional economies for over a century, and social indicators are largely a phenomenon of the postwar world. However, environmental indicators are relatively new and attempt to incorporate the ecosystem into the socio-economic indicators of a study site.

Any interest in defining these indicators primarily comes from the need to monitor performance and to indicate improvements resulting from specific actions. While economists have little difficulty in applying quantitative indicators, sociologists can have great difficulty in creating useful indicators for assessing the quality of life of a social group, as this is an issue that can be approached from different perspectives, many of them intangible. Scientists involved with the environment are considered less likely to have difficulty in establishing practical indicators with which to assess the ecological integrity of an ecosystem, either generally or in specific qualifying aspects.

However, sustainability is more than the interconnection of the economy, society and the environment. It may be something greater and more noble than a dynamic, collective state of grace, a theory such as Gaia (a set of scientific models of the biosphere in which life is postulated that fosters and maintains suitable conditions for itself, affecting the environment), or even the spirit. Instead of asking how can we measure sustainability, it may be more appropriate to ask what degree of sustainability is it?

1.1. The Concept of Sustainability

The concept of sustainability has penetrated most life spheres, not only as a political requirement but also as something that clearly resonates deep within us, even if we have a poor understanding of what it is. The concept first emerged in the mid-1970s, but it exploded on the world stage in 1987 with theBrundtland Report (1987), in which sustainable development was defined as meeting present needs without compromising the ability of future generations to meet their own needs.

Even though this is a very noble goal, this definition challenges interpretation or operational implementation. Most of us would see our personal needs in the context of our circumstances and not as absolute entities. Therefore, our perception of the needs of future generations would impoverish the imagination. "How much is enough?" is a question we have to explore together, but it can only be answered separately. However, we rarely ask this key question individually, let alone collectively.

Once the Earth's ecological integrity is assured and our basic needs are met, how much is enough? The question should be considered in most developed countries, where in the midst of wealth there is still inequality. Increasing inequality is a necessary characteristic for the growth and advancement of an economy. Although it is desirable to achieve a high standard of living, there are finite limits. Our concern for the environment generally decreases with more prosperity, and we should not expect that our pursuit of sustainability should increase as our material wealth increases. Kerala, Cuba, Mennonite and Amish communities are all examples of small societies that practice sustainability, and they all exhibit traits of greater equity, justice and social cohesion.

There are other definitions that ignore human needs and express sustainability in terms of ecological integrity, diversity and limits. However, these definitions also challenge objective interpretation. Such deficiencies in the definitions can cause considerable frustration in a rational way of thinking, particularly for those trying to measure sustainability (Trzyna, 1995). Meanwhile, a reductionist mindset has the ability to link quantitative and productive activity, as in the case of sustainable agriculture, forestry, land management, fisheries, etc. Consequently, growth and sustainable development have been captured as the dominant paradigm. Sustainable development is held up as a new standard for those who really do not want to change the current model of development (Gligo, 1995), and sustainable development alone does not lead to sustainability. In fact, it is possible to support the longevity of an unsustainable path (Yanarella and Levine, 1992). However, the concept is still with us and is becoming stronger.

In general, we have a better understanding of what is unsustainable rather than what is sustainable. Unsustainability is commonly seen as the degradation of the environment (in its broadest sense), the strains of the human population, wealth and green technology in its global limits. Because these effects are entirely of our own construction, their control is, at least in theory, within our capabilities. Human nature tends to promote physical and biological limits towards survival rather than sustainability. We likely think of sustainability in terms of justice, interdependence, sufficiency, choice and above all, (if we were to think deeply about it) the meaning of life.

Sustainability is also non-material life–the intuitive, emotional, and spiritual creativity for those who strive for all forms of learning. Perhaps some truths are really fundamental and universal, if their meaning and spirituality are components of sustainability. These morals and values, however, are not necessarily absolute and can be very difficult to define. For example, values are qualities that are derived from our experiences. If they confirm our default values, then we are more likely to adopt these values. When our experiences are continually at odds with implicit values, we are more prone to change our personal values with respect to the projected values.

Our inability to define sustainability means that we cannot prescribe it. The future may develop according to our vision and ability to always recognize global limits. Sachs (1996) presents three perspectives of sustainable development: the competition implies the prospect that infinite growth is possible over time; the astronaut perspective recognizes that development is poor over time; and the home accepts the prospect of finitude in development. These may be respectively considered as the dominant paradigm perspective, the precautionary principle and conservationist.

Accepting sustainability as a concept can create as many difficulties as the concept of evolution did 150 years ago. During this time, we have not addressed physical consequences involving the collective proficiency requirements for all companies; thus, in general, human awareness has created the concept of ecological crisis with little consequence. Therefore, any discussion of sustainability is essentially a debate about the meaning of what, who, why and how we believe individually and collectively. However, we are very reluctant to participate in the debate on a collective basis, even locally, let alone nationally or globally, in part because it is a messy and time-consuming proposition, i.e., there is a crisis of perception on which one side resides banality, while on the other side there is uncertainty and fear.

1.2. General Indicators

Indicators and measurements are essential components in closed physical systems because they are an integral part of the scientific method. In this context, each indicator must be enclosed between target value limits to guide political and social action. Its usefulness for socio-biophysical closed systems (e.g., human welfare) and, in particular, to open physical systems (e.g., businesses, national economies, regional sustainability) is still unknown because knowledge of the full impact of external factors may not be possible. However, the Earth is ultimately a closed system, except for the flow of energy. In that sense, measures are needed that are theoretically possible globally, but local measures are potentially more meaningful and actionable. The impact of some issues can only be evident at the global level, for example, global warming and ozone depletion, even though the solutions may be local.

Henderson (1991) wrote extensively on indicators, and particularly on current paradigms. The proliferation of indicators is indicative of the confusion and uncertainty of what has been measured as well as the absence of debate and understanding.

1.2.1. Economic Indicators

There is much discontent with economic indicators, as the majority states that they are indicators of something more than the economy. Some do not believe that there are significant economic measures of sustainability.

The most common indicator is the Gross National Product (GNP), now replaced by the Gross Domestic Product (GDP). Daly and Cobb (1994) developed the Index of Sustainable Economic Welfare (ISEW), which recently has been refined as the Genuine Progress Indicator (GPI) by Cobb, et al. (1995). Consumption remains the basis of the index, but instead of adding only negative or harmful consumption (e.g., environmental protection), positive beneficial consumption is also added (for example, voluntary work, child care, housework, etc.). It is difficult to conceive of an index where consumption is the basis for measuring sustainability.

GDP and GPI are aggregations of specific economic indicators, which can be equally sensitive in terms of time, or the actions of adjustment, but which do not apply to social or environmental concerns. Economic indicators are therefore not particularly useful as measures of sustainability, even though economic considerations must be taken into account.

The basis of modern economic theory has a political and a cultural component that addresses scarcity of resources. Affirming the need for a theory that goes beyond that and reflects the basic human needs would be very helpful.

1.2.2. Social Indicators

Overall, there are five types of social indicators: informative, predictive, problem-oriented, program evaluation, and goal set. Several of them are partly economic, environmental and sustainable; they can be combined and compared, such as socio-economic indicators.

Indicators such as the standard of living, which is measured by analyzing time series data on observable phenomena, are called objective. Indicators such as quality of life, which measures the perceptions, feelings and responses through questionnaires with classified scales, are called subjective. The correlation between these conditions is very low, and there are considerable difficulties related to indicator aggregation and the design of weighting schemes.

1.2.3. Environmental / Ecological Indicators

Environmental indicators tend to relate most closely to human activity but may include economic, social and sustainability parameters. Measures may include the quality of living conditions and work environments, including air, land and water, as well as the productive use of resources.

Ecological indicators are more concerned with natural ecosystems; in some cases, human impact is not as obvious. The indicators for the integrity of ecosystems and biodiversity are prominent. The OECD produces a "pressure-state-response" model that many countries have used in the preparation of their "State of the Environment."

Most indicators have thresholds and targets. At present, there seems to be no drive to aggregate indicators or obtain a unique index. However, the Framework Convention on Climate Change (UNFCCC) and the Global Environment Facility (GEF) have very specific defined indicators for the problem of global climate change that may be adopted by all countries. Table 1 provides an overview of each topic in the UNFCCC so that each ecosystem could be described as the physical state of the substances found therein. As indicators for the GEF, they have a direct relationship with the strategic objectives defined by the same agency for action.

TABLE 1. Outline of indicators proposed by the UNFCCC for global climate change.

Topic	Sub-topic	Indicator	Unit of measurement
Atmosphere	Climate change	Emission of greenhouse gases	Gg or ton of CO_{2eq}
	Decreased ozone	Consumption of substances that deplete the ozone layer	Ton of CFC-11 or CFC-12 equivalent
	Air quality	Concentration of air pollutants in urban areas	µg/m³, ppm, ‰
Land Use	Agriculture	Area of arable land under cultivation and permanent	ha
		Fertilizers	kg/m²
		Pesticides	kg/m²
	Forest	Forest area as a percentage of the total area	%
		Intensity of logging	%
	Desertification	Area affected by desertification	km² or %
	Urbanization	Area occupied by informal and formal settlements	km²
Oceans, seas and coasts	Coastal areas	Concentration of algae in coastal waters	mg of chlorophyll/m³
		Percentage of the total population that lives in coastal areas	%
	Fishing	Annual catch of target species	Ton/year
Fresh water	Water distribution	Annual withdrawals of ground and surface water as a percentage of total available water	%
	Water Quality	Biochemical Oxygen Demand (BOD) in water bodies	mg/L
		Concentration of fecal coliform in freshwater	mg/L
Biodiversity	Ecosystem	Area covering selected key ecosystems	km² o ha
		Protected area as a percent of total land area	%
	Species	Abundance of selected key species	# of individuals

1.2.4. Indicators of Sustainability

Sustainability measures today tend to be an amalgam of economic, environmental and social indicators. The first two are susceptible to quantitative measurement because they can be expressed in biophysical terms, while the third is not easily quantified. Therefore, there is a tendency to only view biophysical sustainability.

Examples of sustainability indicators for a city are as follows:

- Per capita income.
- Solid waste generated/water consumption/energy consumption per capita.
- Proportion of workforce at the ten largest employers.
- Number of days of good air quality per year.
- Diversity and population size of particular urban wildlife (especially birds).
- Distance traveled on public transport and private transport per inhabitant.
- Residential density in relation to public space in city centers.
- Hospital admissions for certain types of childhood diseases.
- Percentage of children born with low birth weight.

Boswell (1995) proposed a theoretical basis for sustainable development indicators on a foundation of knowledge in sociology and ecology. Below, we present a set of attributes (energy use, community structure, life history, nutrient cycling, selection pressure and balance) in terms of objectives for the sustainable management of communities. The system lists 23 necessary conditions, but this may not be sufficient. The same author evaluates these goals with the selected sustainable development indicators. While a human ecology approach is clearly appropriate, Boswell (1995) does not recognize that the communities themselves should determine the strategy and indicators.

Whereas these are facets of sustainability, we must look beyond conventional measures to include a sense of quality of life, welfare, relevance, and harmony. We may have to be willing to accept semi-quantitative and qualitative indicators.

Environmental and social indicators are rarely expressed with a unique index. There is some interest in developing a single sustainability index based on a weighting of economic, environmental and social criteria, but this index cannot meet response times ranging from a few years (e.g., medical intervention) to a generation (e.g., global warming).

1.3. Criteria for The Selection of Sustainability Indicators

The monitoring of sustainability is a long-term exercise so it must be flexible. The criteria for selecting appropriate indicators today could be expressed in a straight line with a slope and perhaps a long learning curve, and our ideas and preferences may change over time once complex criteria can achieve amenable results through statistical analysis. Perhaps someone can reduce a large set of indicators into a single sustainability index. Conversely, some communities may prefer or be willing to accept a few qualitative indicators for the sake of simplicity and direct relevance. Excluding qualitative criteria because they are not easily amenable to objective analysis would likely lead to the exclusion of essential characteristics of sustainability.

The numerous sets of criteria, e.g., Liverman (1988) and Seattle (1998), range from the simple (efficiency, fairness, integrity, management skills) to the complex. Hart (1995) believes that the best measures are not yet developed but suggests the following criteria:

- Multi-dimensional, linking two or more categories (e.g., economy and environment).
- Looking to the Future (range 20 to 50 years).
- Emphasis on local wealth, local resources and local needs.
- Emphasis on the levels and types of consumption.
- Measures and visualize changes that are easy to understand.
- Reliable, accurate and updated data available.
- Reflect local sustainability to improve global sustainability.

Social criteria (e.g., quality of life, sense of security, relationship with others) must reflect the degree of choice that a person has in an action. Many of us are locked into our own systems of collective construction within the dominant paradigm (many unsustainable) where the choice of being different can be socially, economically and practically difficult. Examples of this are the use of solar radiation and precipitation in dwellings and foregoing ownership of a car.

1.4. Risk Analysis and Comparative Risk Assessment

In all stages of information, including insufficient quality and quantity or vagueness and uncertainty, where much is at stake and there are several options for action, risk analysis can assist in selecting the most accurate values, lower costs, and/or the lower-risk option. The poorer the information, the greater the uncertainty, and risk analysis may be necessary. We suggest a preliminary stage of data analysis in order to confront a different set of issues and problems with inadequate resources. This technique classifies the problem issues according to the urgency, cost and likelihood of success.

It is frequently argued that there is insufficient or inadequate information to permit taking a rational action, including activities that affect sustainability. However, we know that there are systemic functional weaknesses in both ourselves and in our organizations. Research information actually adds to the uncertainty or controversy; we lose valuable time while more unnecessary work is undertaken. We know the direction that our action should take, but we do not know exactly what that action should be. Many of the problems and solutions are neither technically nor entirely rational. A new methodology for needs that arise may be required for sustainability. They should only be started through social action, where the general population as well as technical experts report on issues and decision-making recommendations.

1.5. Limitations of the Measures of Sustainability

Although we cannot objectively and unambiguously define sustainability, we must not abandon or postpone attempts to measure it. Even if we recognize that there are other equally valid ways of learning, we must begin where we are, even though that may be reductionist, rational or materialistic. We can define the limiting aspects (for example, the sustainable productive capacity of a specific area of the Earth) and trends in the direction of sustainability (for example, increased use of public transport, a more equitable distribution of revenues) and choose indicators that are appropriate and meaningful. The former must be below the threshold of unsustainability. The latter must give directions that require us to act. Many, in fact, are actually indicators of unsustainability. Many discussions and studies on the measurement of sustainability are not defined, nor do they even provide a common understanding of what is measured. The context of sustainability cannot be separated from the measurement.

We recognize at the outset the limitations of quantitative measures. However, we must be on guard to keep the threshold clear. Although sustainability is about quality and other intangible non-physical aspects of life, this does not mean that we are unable to obtain measurements for them. Just as biological indicators (e.g., health of trouts) are now used to measure the quality of industrial effluents alongside conventional physical-chemical indicators, we must be able to obtain parameters that serve us and the Earth.

1.6. Some Indicators to Measure Sustainability

If we know that we are becoming more sustainable without having to measure the "sustainability discourse" as part of the process that then leads to a sustainable lifestyle and measures of it, some of which are relatively easy to measure and some of which are roughly quantified to preset limits. However, if it is consistent, then we can say that achieving sustainability has begun. Therein lay the success of initiatives such as Seattle.

The initial challenge of this discourse is communicating the environmental and social change that is underway within organizations, as groups cannot yet see their particular success as part of the combined progress towards sustainability. The dialogue should be extended to the wider community to open the discussion for a more effective participation on the big issues ahead. Local communities need to renegotiate their sense of community in the modern world and discover new modes of expression.

2. EVALUATION OF SUSTAINABILITY BETWEEN COMBINED CYCLE POWER PLANT PRODUCTION AND A HYBRID SOLAR-COMBINED CYCLE SYSTEM

A discussion on sustainable development must create a process-oriented dialogue and therefore a dynamic concept that establishes priorities; the generic concept of sustainable development must be able to determine its specificity and concreteness at a local and regional level.

This section presents a biophysical, social and economic need for high available renewable resources within our country. As a primary energy source for the generation of electricity in a combined cycle solar energy plant, the sustainability is measured in physical terms, while also taking social, economic and environmental interactions into account.

2.1. Overview

We consider the following questions: What is to be held, for how long, and at what spatial scale? These questions involve social concepts and economic and biophysical factors that should be evaluated as deeply as the scope of this study can allow. This project should also be evaluated superficially, viewing sustainability as a multivariate feature in a socio-environmental system that involves answering additional questions such as: Sustainability for whom, who will carry it out, and how can it be done? Only then can we understand and integrate the plurality of preferences, priorities, perceptions and joint inequalities in the objectives of what is to be held in an appropriate application to the different scales of analysis.

This section requires the evaluation of sustainability between two electrical generation systems: a combined cycle (conventional) and a hybrid solar-combined cycle.

Energy is essential for economic, social and global welfare, but unfortunately, most of it is produced and consumed in unsustainable ways (Yuksel, 2008). The primary source is fossil fuels (oil, coal and natural gas), with more than 90% of global production used to meet commercial energy needs. OPEC forecasts foresee further growth into 2030, both in developed and developing countries. Consequently, energy poverty is a crucial variable for the foreseeable future (OPEC, 2007), while the control of gases and other substances emitted into the atmosphere will become a more urgent matter to be resolved. This condition implies that further improvements must be achieved in the production, transmission, distribution and consumption of electricity (Yuksel, 2008).

In this context, renewable energy sources such as solar, wind, hydro, geothermal and biogas are potential candidates to meet global energy requirements in a sustainable manner. Renewable energy sources have some advantages when compared with fossil fuels (Demirbas, 2000). As a result, the increased use of renewable energy can have a significant environmental effect.

Among renewable sources, solar technologies are attracting worldwide attention (Patlitzianas et al., 2005, Hang et al., 2007); their application in new structures and their adoption in existing ones is currently one of the more common approaches with respect to electricity and heating supply. For example, solar photovoltaic technology worldwide in 2004 reached a production level of 1256 MWp, a 67% increase in production from 2003 (Flamant et al., 2006). Photothermal technology has reached over 430 MW (Morse, 2008).

This trend is expected to continue in the coming years, requiring the creation of specific tools for evaluating the efficiency of solar technology. Classical methods essentially provide tools for the assessment of energy and economy. However, placing these assessments in the broader context of sustainability of the environment require more integrated analyses. From this perspective, one must quantify both environmental and economic costs.

To this end, "emergy" has recently been identified as a valid analysis approach. Emergy can be defined as "useful energy (exergy) of a certain type, which has been used both directly and indirectly in the process of developing a particular product or service" (Odum, 1988; Scienceman and El-Youssef, 1993). Emergy expresses the cost of process-equivalent units of energy, such as solar power. The basic idea is that solar energy becomes the primary unit of energy that expresses the value of any other unit of energy so that it is possible to compare completely different systems, such as Emjoule (emergy joule), also called the emjoule solar, which is designated by the symbol (sej). "Emergy calculations have the same purpose as the Exergy: to capture the energy hidden in the organization and construction of living organisms." It is beyond words to define emerging as "exergy built."

2.2. Definitions

The means used to achieve the desired objectives will be varied, so emphasis should be placed on long-term ecological sustainability. All methods should promote the efficient use of energy and resources, encourage the use of renewable energy sources (and thereby reduce fossil fuel use), reduce costs and increase the efficiency and economic viability of alternative energy sources.

From the environmental point of view, the sustainability of a hybrid solar-combined cycle power generation system will essentially depend on the management and optimization of the following processes:

- Reduction in natural gas consumption; this will also reduce greenhouse gas emissions into the atmosphere.
- Preservation and integration of biodiversity; the use of parabolic trough concentrators requires a large area, so it is important to locate the hub area while affecting as little regional flora and fauna as possible.

Socially, electric power generation should benefit all communities in the area no matter where they are located, thus providing people with energy that can be used to increase productive capacities, self-management and local cooperative mechanisms. One can say that the process is a socially driven activator, improving conditions for all those who receive that energy as well as future generations.

2.3. Systemic Attributes and Operational Definitions of A Sustainable Management System

The following primary schematic characteristics of a sustainable system must be analyzed:

- Productivity: the system's ability to provide the required level of goods and services. Represents the attribute value over a period of time.
- Equity: the system's ability to deliver productivity (benefits and costs) in a fair manner. This implies a distribution of productivity among affected beneficiaries in the present and the future.
- Stability: refers to ownership of the system having a dynamic state of equilibrium. It can maintain the productivity of the system at a level not decreasing over time under normal conditions.
- Resilience: the ability to return to equilibrium or maintain productive potential after the system has suffered major disturbances.
- Reliability: the ability of the system to be maintained at levels close to the usual equilibrium, i.e., temperature shocks.
- Adaptability and flexibility: the ability to find new equilibrium levels to long-term changes in the environment. It is also the ability to actively seek new levels of productivity.
- Self-reliance or self-management: the ability to regulate and control the system through outside interactions. Includes organizational processes and mechanisms of socio-environmental systems to endogenously define their own goals, priorities, identities and values.

It also emphasizes that the sustainability of a system depends on endogenous properties and their external linkages with other systems and structural relationships. These attributes are designed to apply to systems management as a whole, including social, economic, environmental and technological attributes. Focusing on the abovementioned attributes allows for the development of sustainability indicators fundamental to systemic priorities, thereby avoiding long lists of purely descriptive factors and variables.

In operational terms, a sustainable management system will be one that simultaneously allows the following:

- A high level of productivity through efficient and synergistic use of natural and economic resources.
- Reliable production, stable and resilient to major disturbances in the course of time, ensuring access and availability of productive renewable resources; the use, restoration and protection of local resources, proper temporal and spatial diversity of the natural environment and economic activities and risk-sharing mechanisms.
- Adaptability or flexibility to adjust to new conditions of economic and biophysical environment through innovation and learning processes and the use of multiple options.
- Fair and equitable distribution of costs and benefits to the different affected groups, ensuring economic access and cultural acceptance of the proposed systems.
- An acceptable level of self-reliance to respond and externally manage induced changes, maintaining its identity and values.

These five general attributes of sustainability are the basis for the design of indicators:

- Productivity, which can be evaluated by measuring efficiency, achieved average returns and availability of resources.
- Stability, reliability and resilience, which can be evaluated with the trend and variation in the average return, with the quality, conservation and protection of resources, renewability in the use of resources, spatial and temporal diversity systems with a relationship between income and opportunity cost system, and an evolution of jobs created and risk-sharing mechanisms.
- Adaptability, which can assess the range of technically and economically available options, with the ability to change and innovate, strengthening the relationship between the process of learning and training.
- Equity, which can assess the distribution of costs and benefits to participants and target groups, and the degree of "democratization" in the decision-making process.
- Self-reliance, where one can evaluate the forms of participation, organization and control over the system and decision-making.

This project follows the methodology proposed by Masera (1996), which consists of the following:

1. Determining the objectives of the evaluation and defining the management systems to assess their characteristics and the socio-environmental assessment.
2. Selecting the indicators that define the critical points for the sustainability of the system, the diagnostic criteria and the derived sustainability indicators.
3. Measuring and monitoring indicators, including the design of analytical instruments and the procedure used to obtain the desired information.
4. Obtaining and submitting results that compare the sustainability of the analyzed management systems, identify the main obstacles to sustainability, and provide suggestions for improving the system of innovative management.

2.4. Objective of the Assessment: Definition of Management System

Planning to meet the nation's future electricity demand is an issue of paramount importance, considering the urgent need for economic development, the projected population growth and the allocation of capital to finance that growth.

The Mexican Federal Electricity Commission (CFE) forecasts that the demand for electricity will grow 5 to 6% annually in coming years, requiring a dramatic increase in production capacity along with new schematic development. A primary objective for the electricity sector should be a transition from a centralized power system to a geographically distributed and decentralized system, allowing for a wide availability of natural resources in order to focus on the use of renewable energy. In this scheme, we propose a reduction in plant size and geographical dispersion.

In this paper, we carry out a sustainability assessment by comparing a traditional production of electricity through a 316 MW combined cycle plant using natural gas as the primary energy source as well as an innovative system of electricity production through a solar-hybrid combined cycle plant that employs an 80 MW thermal and a 236 MW natural gas energy source.

The large capacity 80 MW solar plant was chosen because of economy of scale (a larger capacity involves occupying a land area of over one million square meters). This capacity is not arbitrary; it is based on an example presented by PURPA (Public Utility Regulatory Policies Act) in the United States of America, which arose from limits for small producers. We selected this capacity to generate the parabolic trough based on the experience and cost information at our disposal. ABB GT24 combined cycle turbines of 236 MW and 316 MW were selected for use in the different alternatives.

The following are the main determinants used to characterize the proposed systems:

2.4.1. Bio-Physical Components of the System

For solar thermal technologies to be effective, the proposed systems must be located in an area with high solar irradiance during most of the year. Sonora State (Northwest of Mexico) was proposed as the construction site of the north plant, as shown in Figure 1.

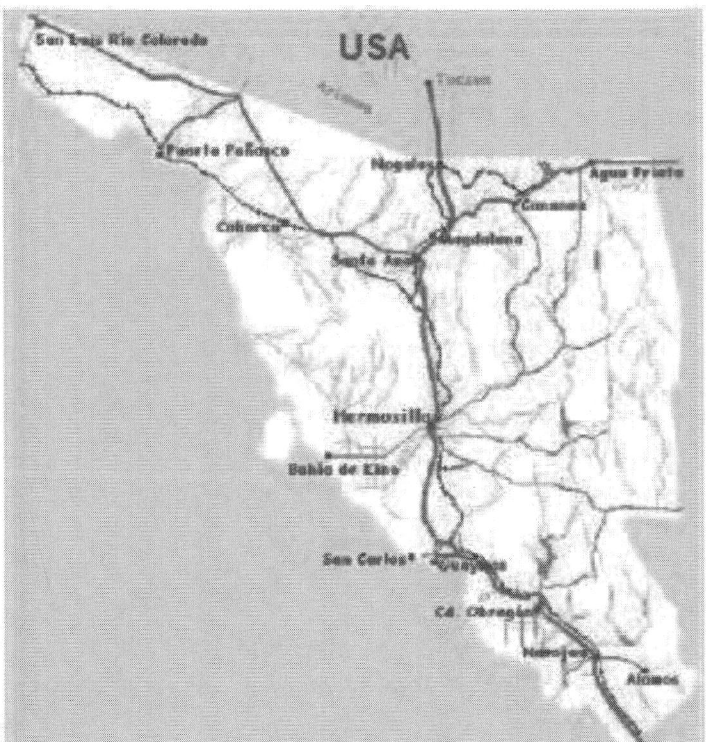

FIGURE 1. State of Sonora (source: IMADES)

Sonora is located within the North West Coastal Plain, which forms a belt 1400 km long, bounded on the east by the Sierra Madre Occidental and on the west by the Gulf of California. It is 250 km wide to the north (Sonora) and 75 km wide to the south (Sinaloa State), with an average elevation of 100 m. The region is mostly flat with a gentle slope towards the sea, interrupted by deeply eroded hills and low mountains or hills surrounded by low lying alluvial plains. From the northern border to the Rio Yaqui, there are large areas of typical desert plains, arreicas and criptorreicas where one can find sand dunes in a half moon.

1. *Climate of the State of Sonora*: The State of Sonora has a dry climate (Figure 2), with an average temperature of 20 °C in the valleys and along the coast, while in the mountain region is 16 °C, with highs of 56 °C and minimum of -10 °C. The northern part of Sonora is characterized by a dry desert climate in the plains near the coast, a temperate rainforest in the mountainous region and the remaining dry steppe. The annual precipitation is 50 to 350 mm in the northwest and 400 to 600 mm in the rest of the state. In the southern desert, the climate is dry and very warm, with a rainfall of 266 mm in the summer.

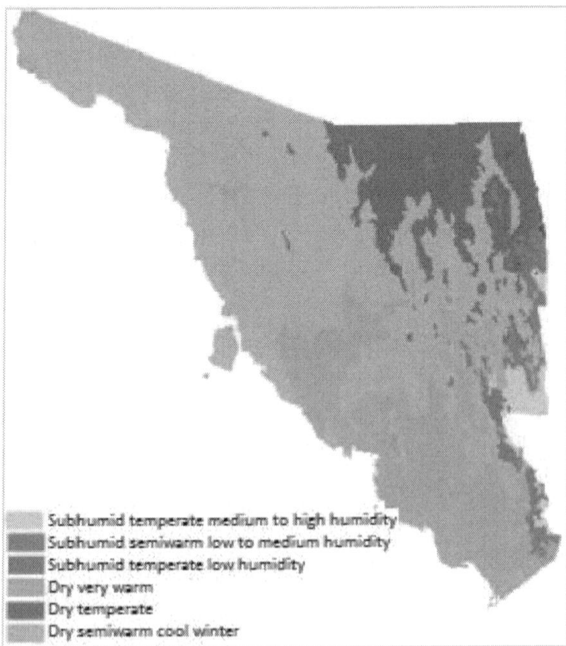

FIGURE 2. Schematic with different climates in the state of Sonora. (Source: IMADES)

1. *Vegetation of the State of Sonora*: Bushes occupy the largest area of the state (38.07%), dominated by ranching and the removal of wildlife for commercial purposes (mesquite, oregano, chiltepin) and crafts (ironwood, etc.). The areas with no apparent use in extension are next in prominence (17.29%). Pastures are predominantly livestock areas and occupy 13.06%. Forests cover 12.57% of the state and are located in the Sierra Madre Occidental; they are characterized as pine-oak, oak-pine and pine, and although there is infrastructure, the forestry operations remain mostly artisanal. Value-mining areas are well distributed in the state (6.28%) and dominated by gold and copper deposits. Approximately 6.01% is suitable for livestock, and the agricultural areas of the state (4.89%) are mostly irrigated. Intensive livestock (poultry, swine, dairy farms and feedlots) occupies a small area (1.15%), although it is economically important.

2. *Soil salinity in the State of Sonora*: Although 90.77% of the territory has no saline problem, approximately 10% is affected by salts at different levels. While 62.3% of the agricultural soils are normal, the rest are either saline-sodic (15.9%), have problems of salinity (12.4%), are strongly saline (3.2%), are strongly saline-sodic (2.3%), have only sodicity problems (1.7%) and are strongly sodic, are moderately saline (0.8%), or are strongly saline (1.4%). Salinity problems primarily exist in the Irrigation Districts of Caborca and the Hermosillo Coast and in the Yaqui and Mayo Valleys. The most important and difficult to eradicate are the saline-sodic soils found in the Yaqui Valley and in the saline delta plain (plains of San Luis Rio, Colorado).

3. *Stationary sources of air pollution*: Using the Information System Rapid Environmental Impact Assessment (SYRIA) simulation model, pig farms, landfills, urban centers, mines, mining and industry emissions were analyzed. In some cases, emission factors were used. The nine municipalities in the State of Sonora comprise nearly 85% of the population and nearly 65% of the productive activities, propagating a proportional burden of pollutants in the atmosphere, which presumably generate 251.2 Mg/year of total particulate matter, 48,037.8 Mg/year of hydrocarbons, and 399.5 Mg/year of carbon oxides in different composition.

4. *Watershed*: The State of Sonora has 12 watersheds, with most domestic consumption taking place in the Sonora River Basin, which passes through the state capital and crosses some of the oldest villages. Next in order of importance are the Yaqui River and Mayo River Basins; they have larger concentrations of people due to an agricultural boom resulting from the construction of hydraulic works. This assertion is reflected in the consumption of water for agricultural activities; water

consumption from the Rio Yaqui and Mayo has increased since the construction of the Alvaro Obregon and Adolfo Ruiz Cortines dams. From the point of view of industrial development, these three hydrologic regions also contribute to the increased water consumption and increased demands on the service sector.

2.4.2. Socioeconomic And Cultural Components

The demand for services and natural resources is determined by the quality of life, which translates to economic growth. To evaluate this demand, we analyzed data on the growth of different economic sectors. The employed population has remained constant over the past three decades. Although the EAP has increased from 25.9% in 1970 to 45% in 1990, the unemployment rate increased from 0.75% in 1980 to 2.5% in 1990. The tertiary sector is the most dynamic in the state, occupying 49%, while 23% are in the primary sector, down from 1970 to 1990. Among the major indigenous groups are the Opata, Yaqui, Papago, Pimas and Seris.

Productive activities (Figure 3) were analyzed based on natural resources, particularly vegetation, as resources show the utilization of the soil. This enables the observation of impacts or consequences of productive activities on the environment.

As mentioned above, the vegetative plane incorporated activities such as aquaculture and was derived from recent satellite imagery. Human settlements and industries were added based on INEGI corrected plans for the 72 largest settlements in the state. Intensive livestock dairies, feedlots, poultry and swine were charted by obtaining the coordinates of each of those registered in the Ministry of Livestock Development, the State Delegation of the Ministry of Agriculture or livestock associations. The mining districts were mapped on the basis of records provided by the Mining Development Division of the Ministry of Economic Development and Productivity.

2.4.3. Technology And Management Components

The state of Sonora has a series of dirt roads, paved roads, highways and railroads linking major cities. As a border state, the highway goes straight to the border in Nogales. There is a pipeline that runs through much of the center of the state, which originates in the U.S., passes by Naco, Cananea, and reaches Hermosillo following the route of the highway. This is a great advantage because this is the area that receives the largest amount of solar radiation in the country. Table 2 compares the proposed systems.

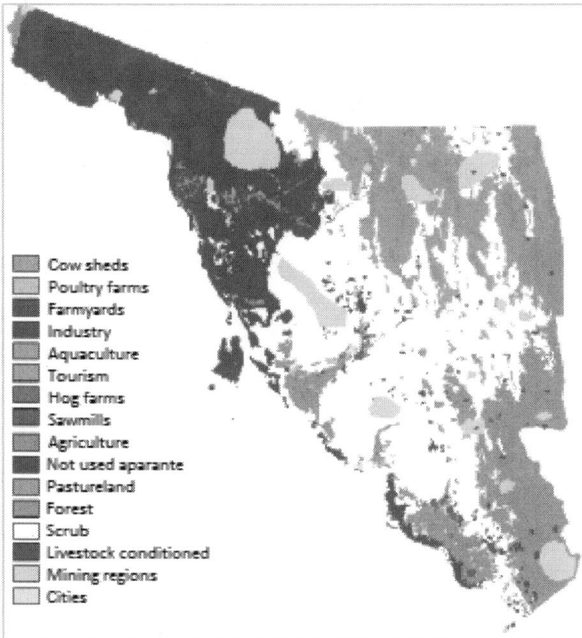

FIGURE 3. Distribution of land use in the State of Sonora. (Source: IMADES)

The plant can be located in any area close to the west of the Hermosillo highway linking the city of Santa Ana south of Nogales. The area around the pipeline from the United States also spans a river, and the amount of solar radiation is the highest in the country. Furthermore, the ground is flat and semiarid.

2.4.4. Identification Of Critical Points In The System

To identify the critical points in the system we ask the following question: What are the environmental factors or processes–technical, social and economic–that individually or in combination may have a crucial effect on the survival of the management system?

a. Environmental aspects: From this point of view, factors that can influence the sustainability of the management system include the following: air pollution from natural gas leaks; the large land area required for the installation of parabolic trough concentrators; large loss of cooling water for weather at certain times of year; low yields from cloudy weather; emissions from burning natural gas; change of land use; erosion; etc.

TABLE 2. Comparison of the characteristics of each system.

System Determinants	Traditional System	Hybrid System
Generation Capacity	300 MW combined cycle	50 MW solar thermal (PT) 250 MW combined cycle
Primary energy used	natural gas (imported)	solar energy and natural gas (imported)
Gross efficiency of c.c.	46.79%	46.79%
Net efficiency of c.c.	45.38%	45.38%
DSG system net efficiency with Parabolic trough	-----	23%
Generation time with natural gas	5694 h/year (p.f.= 0.65)	5694 h/year (p.f.= 0.65 for c.c.) 13.2 h/day (85.3%)
Generation time with solar power	-----	2445 h/year* 6.7 h/day (14.7%)
Solar radiation design	-----	2772 kWh/m2
Maximum solar radiation	-----	3122 kWh/m2
Natural gas heat value	9200.14 kcal/m3	9200.14 kcal/m3
Solar concentration area	-----	570 265 m2
Required total area	846 476 m2	1 316 741 m2
Life	30 years	30 years
Domestics inputs	37.40%	53.70%
Imported inputs	62.60%	46.30%
Funding	CFE and private investment	CFE, World Bank,private investment and GEF

[i] - * The combined cycle always works and the plant is at full capacity for only 6.7 hours per day.

[ii] - Where: c.c. --- Combined cycle p.f. --- Plant factor (taken from COPAR) P.T. --- Parabolic trough

a. Socioeconomic aspects: These aspects are highly dependent on electricity prices because instability will affect the entire future of a plant using imported natural gas. Though this natural gas pipeline is national, it remains very sensitive to the unit price. International borrowing may be required to finance the construction of any system, and the construction may entail in a high migration of population for the construction of the plant, resulting in an imbalance in the surrounding communities. Combined cycle technology requires a great deal of imported equipment and will be subject to prices quoted in dollars or euros. The cost of labor will be slightly higher compared to the rest of the country, as this is a region near the border. Additionally, the use of an alternate source of energy can create suspicion among investors. Recently, CFE tender-type parabolic trough plants have been deserted both in Agua Prieta and Puerto Libertad (Sonora) due to administrative – not technical– reasons.

2.5. Selection of Indicators

To define the indicators used in this evaluation, we must select those that are inclusive, i.e., those that describe rather than analyze processes. The indicators must be easy to measure, easy to obtain and be appropriate for the system under analysis. They must be applicable in a defined range of ecological, socioeconomic and cultural conditions and have a high level of reliability. These indicators should be easy to understand for most readers and be able to measure changes in system characteristics over time in a practical and clear manner. Finally, the measurements must be repeatable over time.

For this paper, we will consider three areas of evaluation: economic, social and technical/environmental, placing the general attributes of sustainability in each of the areas proposed by their own diagnostic criteria.

2.5.1. Economic Indicators

To select these indicators, we must first state the diagnostic criteria to follow and then the indicator to use for defining the general attributes of sustainability.

- Productivity: You can assess profitability and efficiency indicators by using Net Present Value, Internal Rate of Return and Cost/Benefit. Other indicators may include the investment cost, turnaround time, etc.
- Stability, Resilience and Reliability: We can assess the diversification of fuel use and risk measurement mechanisms, using indicators for credit, insurance, leverage, percentage of income derived from the use of different primary energies, price of natural gas, etc.
- Adaptability: We can evaluate the options of primary energy use and technology options by indicating of number and type of primary energy options and technologies available, the cost at low loads, cost of generation on cloudy days, low demands, etc.

- Equity: The diagnostic criterion is the adaptability of technology and employment trends using indicators such as the cost of investment/production revenue, number of jobs created (temporary and permanent), access to fire insurance, etc.
- Self-Reliance: The indicators measure the level of self-financing, the degree of indebtedness, domestic savings, percentage of self-produced energy use, etc.

2.5.2. Technical and Environmental Indicators

These indicators give us information about the ability of the proposed systems to be environmentally "productive." Sustainability must sometimes include indicators describing the state of the environment or the processes of prevention and protection of environmental degradation.

- Stability, Resilience and Reliability: Can be used as an indicator of land use patterns, number of species in the area, soil quality and water, soil degradation, disasters, climate change, soil chemical properties, physical soil properties, distribution of natural capital in each region, and so on.
- Self-Reliance: This requires indicators of energy subsidy, energy efficiency and degrees of external dependence.

2.5.3. Social Indicators

This type of indicator is very difficult to quantify, especially for a production plant that will supply power to the communities and surrounding cities in Sonora. A much larger study is required to determine the exact number of beneficiaries and the investment schemes to be used for construction and operation. Some of the indicators suggested in the literature (Masera, 1996) are as follows:

- Equity: The distribution of benefits can be used as an indicator of the number and type of benefits by gender, social sector, age, ethnicity, etc. The factors influencing decision-making may include policies, group's resistance, lobbyists and others.
- Stability, Resilience and Reliability: The ability to overcome serious events can affect the survival of the project after conflicts, problems or lack of financing. The processes of learning and training can reference the type and frequency of training, knowledge sharing mechanisms between members, etc.
- Adaptability: Human resource development can be evaluated using indicators such as concepts, methodology and ownership by the

community as a capacity for change. We can assess changes in objectives, projects, personnel, and adaptation to changes in the different aspects of production, etc.

- Self-Reliance: Participation is evaluated by the number and frequency of the different phases of the project. We measure the power control that decides on critical aspects of the organization with respect to the type, structure and permanence of the organization.

2.6. Measurement and Monitoring Indicators

The above indicators will be evaluated quantitatively or qualitatively, depending on the question, because some of them will be justified by arguments or theoretical reasons, partly due to the difficulty in assigning a number to an assessment of non-numeric type. First, we will present the economic indicators, followed by the technical-environmental indicators and then social indicators.

TABLE 3. Economic variables of each model (Source: own data).

Indicator	Conventional System	Hybrid System
Generation [GWh]	1,708	1,708
Investment cost [USD]	$135,000,000.00	$237,316,931.13
Fuel cost [USD]	$68,505,647.32	$57,088,039.43
Operation and Maintenance Cost [USD]	$8,634,951.00	$8,334,592.50
Unit Cost of Generation [USD/kWh]	$0.05	$0.04
Unit Cost of Investment [USD/kWh]	$0.08	$0.14
Internal Rate of Recovery	26.90%	19.60%
Net Present Value [USD]	$38,943,485.89	$33,155,291.46
Annuity equivalent [USD]	$7,821,505.70	$6,659,196.56
Benefit/Cost	4.11	2.61
Recovery period	4.8 years	3.1 years

2.6.1. Economic Indicators

Regardless of the general attributes of evaluating sustainability indicators, they are calculated individually to reach a conclusion. Table 3 shows the results of these calculations (Geyer, et al., 2004).

The economic and financial analysis necessary to reach these results was performed on a spreadsheet, considering each year of construction, testing, operational development and economic variables.

For economic indicators, we used data from the Costs and Benchmarks for Formulation of Investment Projects in the Electricity Sector - Generation (CFE, 2007) for the combined cycle units, while the thermal data were taken from Hertlein et al. (1990) and Franz Trieb (2009).

2.6.2. Technical and Environmental Indicators

The location of the proposed plant is 73 km northwest of City of Caborca, off Highway 37 (60 km in a straight line). The vegetation of the area consists of scrub and grassland as non-endemic species. The fauna consists primarily of rodents, reptiles and insects. No crops are grown in the area.

The degradation phenomena studied consisted of erosion, salinity and pollution (soil, water and air). In most cases, estimates were made in the absence of available information by using mathematical models with the aid of GIS and satellite imagery to update information. The erosion in the area lies between 4 and 10 ton/ha. In terms of salinity, the area is within the affected soils. There are no landfills in the vicinity.

The selected region has few clouds for most of the year. However, as the present electricity generation derives from natural gas, the combined cycle plant will operate continuously and the total capacity of the plant will be operational within the CSP.

The energy subsidy should be completely designated as external and not just for the region but also for the nation, as the pipeline that feeds the plant comes from the United States. The energy efficiency is the highest in Mexico. The total conversion efficiency of the solar thermal power plant varies from 21 to 23%, and it can be significantly improved if a Direct Steam Generation solar field is used to deliver steam at 550 °C and 100 bar (Zarza, 2004). The CO_2 emissions from each of the proposed systems are shown in Table 4.

TABLE 4. Atmosphere emissions for each model.

Indicator	Conventional System	Hybrid System
Amount of fuel	386 181 018 m^3/year	337 125 816 m^3/year
Amount of CO_2emitted	275 844 ton/year	240 804 ton/year

2.6.3. Social Indicators

The indigenous groups in Sonora who live around the area proposed for the construction of the plant (the Opata, Yaqui, Papago and Pimas) subsist primarily through activities such as the manufacture of handicrafts, animal husbandry and subsistence farming. The construction of the plant would mean a source of temporary employment for them, bringing benefits in both economy and quality of life.

The primary operator will be CFE, although it is very likely that contractors will decide less important aspects that may have a major impact on the region.

In the event of any social conflict in the area, the construction phase will have to stop for security reasons. However, if the pipeline continues to provide natural gas, then the plant will continue producing. There is sufficient security in the area to guard against rebel groups tampering with transmission towers or the pipeline.

Staff at the plant must be local people who receive sufficient training in all aspects of the plant; this can provide a continuity of (very general) knowledge to the community.

The organization of the plant will likely come from CFE because it is the institution that controls and manages the production of electricity. They already have defined organizational schemes in place for the initial production of a new power plant.

The beneficiary communities surrounding the hybrid plant, including the cities of Hermosillo and Santa Ana, are not intended to affect social stability during the construction period (1 to 3 years) and afterwards during operation.

2.7. Evaluation Results

The objective of the proposed hybrid system is the generation of electricity for the area by installing the latest technology, meaning that the technology used has the highest possible conversion efficiency of primary energy, that the solid and liquid waste emissions are minimized, and that the system incorporates the additional use of a renewable energy source. According to calculations above, the obtained results show that the amount of generated CO_2 is less than that emitted by the conventional system; hence, the risk for environmental pollution is reduced for future generations. The plant is also adapted to an area with poor socio-ecological circumstances, which will dramatically improve the benefits for future generations.

Production structures for generation, distribution and consumption will provide electrification services and reliable energy necessary for the progress of the region, which facilitates total employment and meaningful work, thereby improving human capabilities for the inhabitants of the region.

With the launch of a hybrid power plant (solar combined cycle), low resource consumption technology is developed that adapts to local socio-ecological circumstances. Because the primary energy sources are low-polluting solar energy and natural gas, there are still significant risks for the present and future. Increasing the electrical infrastructure in the region by consuming imported gas will not preserve this resource for future generations for other uses. It will, however, conserve resources in the zone if outside resources are used.

We must ensure the satisfaction of some of the most basic human needs, such as the provision of high-quality energy. By implementing an innovative system of this type, we promote cultural diversity and pluralism; by using new commercial technology, we can share experiences with other international institutions on the construction of field parabolic trough concentrators and on the development of the plant and its operation.

This should help to reduce the aspects that make it less sustainable to allow the use of more solar energy as renewable primary energy.

In terms of economic indicators such as initial investment, it is slightly more expensive to implement a hybrid plant. However, costs would be absorbed by international organizations and the CFE, whose resources are governmental and therefore contributed by people from across the country. Despite this, the remaining features and sustainability objectives are met; therefore, it can be said that a sustainable system has a very promising future. The advantages outweigh the disadvantages, and thus it is worth implementing a solar power plant for generating electricity under the study conditions. It is a sustainable project from a technical, ecological and social standpoint, but from an economic point of view, it is not entirely sustainable because it requires external resources, which does not comply with the characteristics of self-sufficiency.

As a complement to the results of this analysis, there is an internal report from the Institute of Engineering (Almanza, et al. 1990) that offers a more formal and technical evaluation of the climatic conditions of the proposed area.

3. CONCLUSIONS

The amount of CO_2 emission from the hybrid system is less than the conventional system. Production structures are generated, and the plant develops hybrid technology with a low consumption of resources, which are adapted to local socio-ecological circumstances.

Since there is a wide acceptance in all social sectors of the concept of sustainability, the proposal made in this paper aims to ensure a sustainable supply of high quality electricity to a region where one of its main natural resource is the Sun spite that, today, still we can not objectively define that term and therefore implement it.

It is very important to make clear that sustainability goes beyond ensuring the environmental integrity of a site and the standard of living of a population, should address the concept of "quality of life" and a form of collective life.

Sustainability is now in a further process of discourse, and efforts to measure it should become a state priority. Institutional initiatives and debates about the measurement of sustainability in general show resistance in committing to this concept. Therefore, there is no common shared understanding of what has been measured.

Sustainability indicators are often an amalgam of economic, social and environmental indicators, but recently, they are showing signs of maturity with better measures of sustainability. These indicators, however, are limited and may reflect unsustainable measures. Their primary value is to indicate the direction of change rather than a swing state.

The indicators are just the initial map and not what could be called the territory. The difficult task of achieving sustainability is another issue.

In consulting references, it follows that the most successful initiatives in measuring sustainability are those initiated and controlled by autonomous public groups (e.g., Sustainable Seattle 1998), where the process is more important than the indicators.

The greater the public involvement in the execution of a community role (for example, consensus conferences, citizen juries, etc.), the more likely we are to achieve sustainability.

We need to address the fundamental existential questions and find meaning in life if we are to achieve sustainability.

The emergy evaluation assigns a value to products and services through their conversion into an equivalent form of energy: solar energy (Odum, 1983, 1996).

Solar energy is used as the common denominator through which different types of resources, whether energy or material, can be measured and compared with others.

REFERENCES

1. Almanza, R, Valdés, A, Lugo, R, Zamora, J M, Sandoval, C y Estrada Cajigal, V. (1990). *Estudio del comportamiento de sistemas solares térmicos para generar electricidad en 5 comunidades de Sonora o Baja California.* Proyecto 0102 Instituto de Ingeniería, UNAM. Patrocinado por CFE. (abril 1990) pp. (1-254)
2. Boswell, M.R. (1995). Establishing indicators of sustainable development. *Proceedings of the Annual Conference of the Association of Collegiate Schools of Planning*. Detroit, Michigan, USA. October 1995.
3. Brundtland, G. (Chairman). (March, 1987). Report of the World Commission on Environment and Development: *Our Common Future.* Available in: http://www.un-documents.net/wced-ocf.htm
4. Cobb, C., Halstead, T. and Rowe, J. (1995). If the GDP is up, why is America down?, In: *Atlantic Monthly digital edition,* Available from: http://www.theatlantic.com/past /politics/ecbig/gdp.htm
5. Comisión Federal de Electricidad (CFE). (2007). Costos y Parámetros de Referencia para la Formulación de Proyectos de Inversión en el Sector Eléctrico – Generación. (April 2007), México.
6. Daly, H. and Cobb, J. (1994) *For the Common Good: Redirecting the Economy toward Community, the Environment, and a Sustainable Future.* (2nd edition). Beacon Press. ISBN 9780807047057. USA.
7. Demirbas, A. (2000). Biomass resources for energy and chemical industry. In: *Energy Education, Science and Technology.* Vol. 5, pp. (21–45)

8. Flamant, G., Kurtcuoglu, V., Murray, J., Steinfeld, A. (2006). Purification of metallurgical grade silicon by a solar process. *Solar Energy Materials & Solar Cells*. Vol. 90, No. 14, (September 2006), pp. (2099–2106), ISSN 0927-0248.
9. Gligo, N. (1995), Situación y perspectivas ambientales en AméricaLatina y el Caribe, *Revista de la CEPAL*, No 55 (LC/G.1858-P), Santiago de Chile.
10. Griffin, J. (Ed.) (2007). *World oil outlook 2007*. Organization of t Petroleum Exporting Countries, ISBN: 9783200009653, Vienna, Austria.
11. Hang, Q., Jun, Z., Xiao, Y., Junkui, C. (2008). Prospect of concentrating solar power in China the sustainable future. *Renewable & Sustainable Energy Reviews*. Vol. 12, No. 9, (December 2008), pp. (2505–2514), ISSN: 1364-0321.
12. Hart, M. (1995). Guide to Sustainable Community Indicators. Ipswich, Maine. QLF/Atlantic Center for the Environment.
13. Henderson, H. (1995). The Indicators Crisis, In: *Paradigms in Progress: Life Beyond Economics,* pp. (147-192), Berrett-Koehler Publishers, ISBN 978-1881052746, USA.
14. Henderzon, H. (1991). Paradigs in progress: Life beyond economics. (2nd Ed.). Knowledge Systems. ISBN 9780941705219. USA.
15. Hertlein, Klaiss, and Nitsch. (1990). *Cost analysis of solar power plants*. American Solar Energy Society. USA.
16. Holden, M. (2006). Sustainable Seattle 1998: The Case of the Prototype Sustainability Indicators Project. In: *Community Quality-of-Life Indicators Social Indicators Research Series.* (2006). Vol. 28, pp. (177-201), DOI: 10.1007/978-1-4020-4625-4_7
17. Liverman, D, Hanson, M E, Brown, B J, y Meridith, R W. (1988). Global sustainability: towards measurement. In: *Journal of Environmental Management*. Vol. 12, No. 2, (1988), pp. 133-143, DOI: 10.1007/BF01873382
18. Masera, O.; Astier, M. (1996). *Metodología para la Evaluación de Sistemas de Manejo Incorporando Indicadores de Sustentabilidad.* Grupo Interdisciplinario de Tecnología Rural Aplicada, México.

19. Morse, F. (2008). Concentrating Solar Power (CSP) as an option to replace coal. In: Energy & Climate Mini-Workshop, Washington, DC. 14/05/2012. Available from: http://ossfoundation.us/projects/energy/2008
20. Odum, H.T. (1988). Self-Organization, Transformity, and Information. *Science*, Vol. 242, No. 4882, (November 1988), pp. (1132-1139), ISSN: 0036-8075 (print), 1095-9203 (online).
21. Organization of the Petroleum Exporting Countries (OPEC). (2007). *World oil outlook 2007.* Technical Report OPEC. ISBN: 9783200009653, Vienna, Austria.
22. Patlitzianas, K., Kagiannas, A., Askounis, D., Psarras, J. (2005). The policy perspective for RES development in the new member states of the EU.*Renewable Energy*. Vol. 30, No. 4, (April 2005), pp. (477–492), ISSN: 0960-1481.
23. Rees, W., Wackernagel, M., and Testemale, P. (1998). *Our Ecological Footprint: Reducing Human Impact on the Earth*. New Society Publishers, ISBN 9780865713123, Canada.
24. Sachs, W. (1996). What kind of sustainability?, *Resurgence*, No 180, (1996), pp. (20-22).
25. Scienceman, D.M. and El-Youssef, B.M. (1993). The System of Emergy Units. *Proceedings of the XXXVII annual meeting of the International Society for the Systems Sciences*. New South Wales, Australia, July 1993.
26. Trieb, F., Schillings, C., O'Sullivan, M., Pregger, T. and Hoyer, C. (2009). Global Potenctial of Concentrating Solar Power. *Proceeding of the 15th Solar PACES International Symposium on Concentrated Solar Power and Chemical Energy Technologies.* Berlin, Germany, September 2009.
27. Trzyna, T. (1995). A Sustainable World; Definig and measuring Sustainable Development. IUCN. Earthscan.
28. Trzyna, T. (Ed). (1996). *A Sustainable World: Defining and Measuring Sustainability*. International Union for Conservation of Nature, ISBN: 978-1880028025, Switzerland.
29. World Commission on Environment and Development. (1987). *Our Common Future*. Oxford University Press. ISBN 9780192820808. Great Britain.

30. Yanarella, E.J. and Levine, R.S. (1992). Does sustainable development lead to sustainability?, *Futures.* Vol. 24, No. 8, (October 1992), pp. (759-774), ISSN: 0016-3287.
31. Yüksel, I. (2008). Global warming and renewable energy sources for sustainable development in Turkey. *Renewable Energy*. Vol. 33, No. 4, (April 2008), pp. (802–812), ISSN: 0960-1481.
32. Zarza, E., Valenzuela, L., León, J., Hennecke, K., Eck, M., Weywers, H., Eickhoff, M. (2004) Direct Steam Generation in parabolic troughs, final results and of the DISS project. *Proceeding of the 12th Solar PACES International Symposium on Concentrated Solar Power and Chemical Energy Technologies.* Oaxaca, México, October 2004.

9 CHAPTER

EVALUATION OF RENEWABLE ENERGY VULNERABILITY TO CLIMATE CHANGE IN BRAZIL: A CASE STUDY OF BIOFUELS AND SOLAR ENERGY

Antonio Oscar Jr., Wanderson Luiz Silva, Vera Ruffato, Renata Barreto, Marcos Freitas

Federal University of Rio de Janeiro, Rio de Janeiro, Brazil

ABSTRACT

This study aims to calculate indicators and indexes to subsidize the analysis of vulnerability and adaptation of the renewable energy sector to climate change in Brazil, focusing on biofuels and solar energy. For biofuels, in general, the Brazilian coast will be a propitious area for agricultural productivity during the XXI century, but these are areas historically intended for occupation and development of the urbanization process, that is, with limited land availability and supply for primary production. In some parts of the Northeast, Midwest and South of the country, offer for the cultivation land will be reduced. For the solar energy is observed that Brazil has area and highly expressive power for the use of this power, both today and in the coming decades, especially in the North, Northeast and Midwest. In statistical terms, the Mann-Kendal test and Sen's Bend point to a very weak tendency to useful radiation indicator in all regions of Brazil by 2100.

In addition, it is projected a significant increase in mean air temperature by the end of XXI century across the country that can mean a reduction in power conversion capability, which is sensitive to ambient temperature variations, especially in the Midwest and North of the country.

Keywords: Renewable Energy, Climate Change, Brazil, Biofuels, Solar Energy

1. INTRODUCTION

Climate change increasingly occupies an important position within the various sectors of society, both nationally and internationally. Several studies, including the fifth assessment report of the Intergovernmental Panel on Climate Change [1] , indicate that the growing trend of greenhouse gases is the main agent of such changes. Thus, there is an imminent concern about the mitigation of these future climate changes, making it necessary to establish measures to adapt to possible impacts.

As the power generation and consumption are key to socio-economic development, it is essential to understand how the energy production infrastructure responds to the clash of climate change. From this knowledge is possible to measure the vulnerability of a nation and promote development of alternative means of it, offering appropriate prospects for the future. In Brazil, due to its characteristic of primary producer associated with continental proportions of its territory and climatic aspects, its energy matrix were guided primarily in renewable energy.

Energy sources exploited in Brazil include hydropower, thermal power, solar power, biofuels and wind energy. According to the National Electric Energy Agency [2] , the country has 3336 projects of electricity generation in operation, representing an output of approximately 118 Gigawatts. Thus, significant climate changes may represent substantial impacts for development and for the country economy in this sector. Specifically this research, the main focus will be on solar energy and biofuels, as these sources have significant enhancement conditions and vulnerability to present and future climate scenarios.

The energy originated from biofuels is from renewable sources which in Brazil are essentially represented by biodiesel and ethanol. For proper use of land with purposes of exploitation of agricultural potential for the generation of biofuels, in addition to economic and social factors, it must take into account basic environmental factors, considering the physicochemical requirements of each culture, as well as the ecology of the region. Among the physical and ecological conditions, the climate plays an important role, especially with regard to water availability, which depends directly on rainfall supply and intensity of evapotranspiration process, elements that act as limiter to vegetation.

For its intertropical location, Brazil has a high incidence of solar radiation throughout the year, making this a potential source for the exploration of energy in order to supply the increasing demand of energy in the country. Most of Brazil is located relatively close to the equator, so that there are not wide variations in daily sunlight. However, most of the population and socio-economic activities of the country are concentrated in more distant regions of equator, as in the Southeast and South.

In view of the above perspectives, this work presents a vulnerability analysis and interpretation of the impacts (both positive and negative) of future climate changes in the renewable energy sector of biofuels and solar energy, in order to support researches that aim to adaptations and implementations in the current Brazilian energy matrix.

2. METODOLOGY

2.1. Data from Climate Model

The Numerical Models of the Earth System (NMESs) are the most promising tools to develop projections of future climate change. Through the simulation of important physical and dynamical processes, NMESs can represent the complex nonlinear interactions that influence the climate as well as the interactions between the components of the Earth system (mainly atmosphere, biosphere and hydrosphere) and feedback mechanisms, including changes in the frequency and intensity of extreme events [3] . In addition, NMESs can simulate future climates in response to changes in the concentration of greenhouse gases and aerosols.

For this study, the regional climate was simulated by Eta model of Weather Forecasting and Climate Studies Center/National Institute for Space Research (CPTEC/INPE), which is coming from the original Eta model [4] developed at the University of Belgrade (Serbia) and operationally implemented by the National Center for Environmental Predictions (NCEP) in the United States. The Eta 20 km regional model scenarios were simulated using the global model boundary conditions MIROC (Model for Interdisciplinary Research on Climate), a model developed in cooperation with the climate research center in Japan. From these integrations, the model is called Eta-Miroc5 [5] [6] .

In this study, the average of thirty years is used as reference model for the processing of data and evaluation of future scenarios of the indicators described hereafter. The RCP 8.5 scenario [1] was selected to perform the climate change analyzes. Historical values (1961-1990) and projected future values (2011-2040, 2041-2070 and 2071-2100) were calculated. Then the future-present differences were computed in order to verify the simulations of future trends.

2.2. Biofuels Indicator

Assuming that water availability is fundamental to the farm of genres used for the production of biofuels, including ethanol, it is used the aridity index (D) proposed by [7] :

$$D = \frac{1}{(0.01 * IM) + 1}$$

where $IM = 100 * \left(\frac{PRECP}{EVTP} - 1 \right)$ is the Thorntwaite Moisture Index, PRECP is the precipitation and EVTP is the evapotranspiration.

From this method, it was shown that occurrence of areas of desertification more emphatic process, so those with very low agricultural potential, are in the range 2 < D < 7.

To complement this methodology is also used as reference humidity data in the soil in 0.1 m, which aims to support the assessment of water availability to the root system of plants. Furthermore, air temperature is also analyzed, from which proposes an index based on its variation (projected temperature-historical temperature), allowing estimation of the thermal stress of the plant in climate change scenarios, fundamental factor that directly affects the photosynthetic capacity of the plant.

2.3. Solar Energy Indicator

Assuming that the use of solar energy depends on the incident solar radiation in a given area, an efficient method of analyzing the view of the offer of this source is the radiation balance in a given region, which can be determined by calculating the balance of radiation. This radiation (Rn) is the available energy in the earth-atmos- phere system and is defined as the radiation balance of all radiative fluxes leading to and from a surface [8] . Therefore, to define a useful radiation indicator (Ru) is required equating the variables of shortwave radiation incident to the surface (OCIS), albedo (α), short wave radiation emerging at the top of the atmosphere (ROCE) and short wave radiation emerging to the surface (OCES). Thus:

$$\mathrm{Ru} = \mathrm{OCIS} + \mathrm{ROCE} - \alpha \cdot \mathrm{OCIS} - \mathrm{OCES} \left[\mathrm{W/m^2} \right]$$

The albedo (fraction of incident radiation which is reflected) regarded was the planet mean value, so 0.30 or 30%. The maps show the average annual values of the indicator for a particular day of the year in Brazil, so Ruis multiplied by 24 hours, having a send unit $W \cdot h/m^2 \cdot day$.

The equipment controllers of solar energy can be markedly influenced by variations in ambient temperature. Thus, also it is performed an analysis of projections of future increase or decrease in air temperature along the XXI century in Brazil.

2.4. Statistical Tests

For the evaluation of significance in climatic trends, it is used the Mann-Kendall nonparametric statistical test [9] . This test has been the most appropriate method to analyze the significance of possible climate change in time series [10] .

Sen's Bendis also calculated in this work. This is a non-parametric method (assuming a linear trend) used to estimate the trends magnitude [11] . As Sen's Bend is insensitive to outliers and missing data, such testis more rigorous than the usual curvature of regression and provides a more realistic measure of trends in a time series.

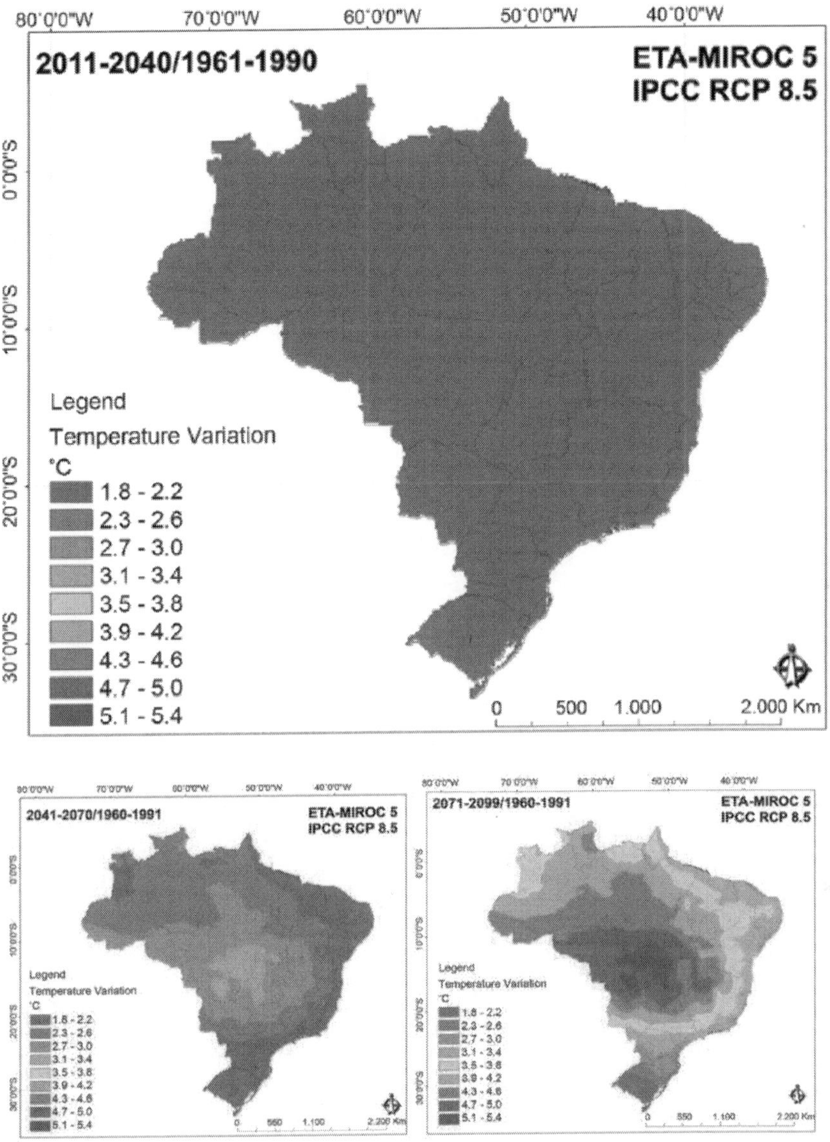

Figure 1. Air temperature differences between the future climate (2011-2040, 2041-2070 and 2071-2100) and the present climate (1961-1990) projected to Brazil through Eta-Miroc 5 regional climate model in the IPCCRCP 8.5 scenario.

Table 1. Statistical analysis of air temperature at 2 meters for Brazilian regions. Values in red were considered significant at a confidence level of 90%.

Region	Mann-Kendall	Sen's Bend
North	0.09	+1.37°C (by slice)
Northeast	0.09	+1.22°C (by slice)
Midwest	0.09	+1.51°C (by slice)
Southeast	0.09	+1.27°C (by slice)
South	0.09	+0.87°C (by slice)

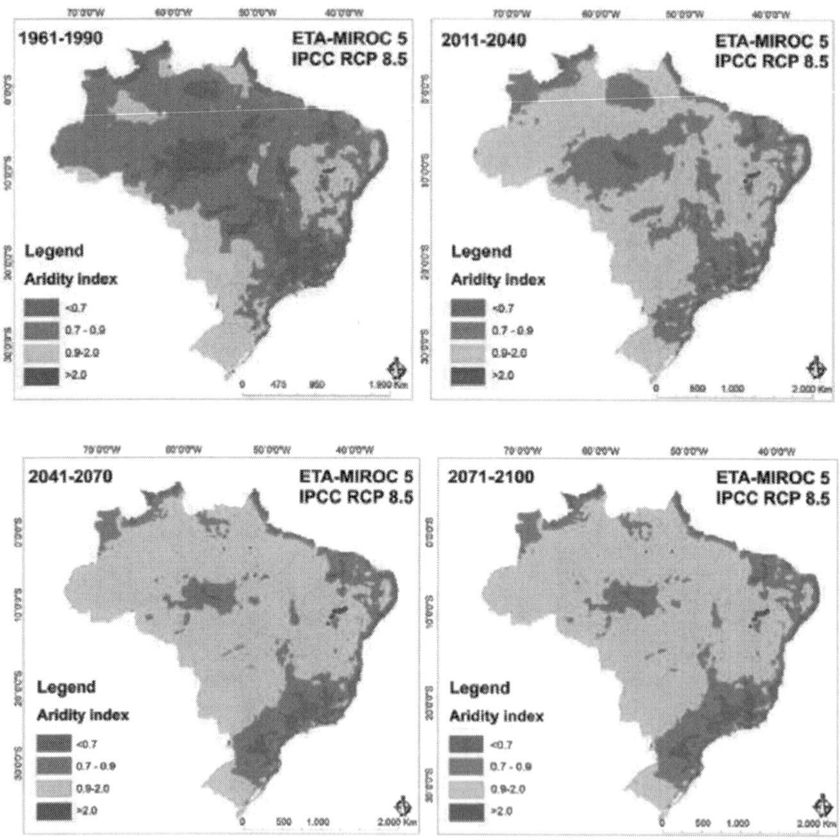

Figure 2. Present climate (1961-1990) and simulated future climate (2011-2040, 2041-2070 and 2071-2100) projected for the aridity index in Brazil through Eta-Miroc5 regional climate model in the IPCCRCP 8.5 scenario.

3. RESULTS

3.1. Air Temperature

Figure 1 shows the differences in air temperature between the future climate (2011-2040, 2041-2070 and 2071- 2100) and the historical period (1961-1990) projected to Brazil through Eta-Miroc5 model in IPCC RCP 8.5 scenario. The estimates show an increase of about +2.0°C in Brazil during 2011-2040. During 2041-2070, the projections indicate a similar temperature rise, but more sharply in the center of the country, with values reaching up to +3.0°C in the Midwest region and vicinity. Analyzing the period 2071-2100, it is noted that the increase in air temperature is emphatic in Brazil. The simulations show an increase between +2.0°C and +3.0°C in the South and in the east coast, while in up-country the difference can reach up to +5.5°C in the Midwest and something around +4.0°C in the North and in the most part of the Northeast and Southeast regions.

In statistical terms, the analyzes indicate a significant upward trend in the average air temperature at 2 m for the five Brazilian regions by the end of the century, confirming the possibility of warming and frequency and intensity growth of temperature extreme events (Table 1). Based on this analysis, the Midwest and North Regions Brazil present the highest rates of average temperature rise, with +1.51°C by slice of time and +1.37°C by slice of time, respectively. The region with the lowest magnitude of increase in the period analyzed is the South, with an increase of approximately +0.87°C by slice of time.

3.2. Biofuels (Bioethanol)

The aridity index, formulated in the previous session, is an excellent indicator to assess the water availability in a particular climate type, pointing regions susceptible to the occurrence of severe droughts and desertification process. In the Brazilian, case is notoriously the prone to dryness of the Wild Northeast. From the proposed index (Figure 2), the northern state of Bahia is presented as area of desertification process, which evolves in terms of extension from the time slice 1961-1990 to 2071-2100. It also arise desertification outbreaks in the State of Pará.

The coast of Brazil, both in its east coast as northeast remains with high moisture index, as well as the southeast of the country, which suffers with significant retraction of this area in 2011-2040 period, particularly in the States of São Paulo and Paraná, but from 2041 onwards, it rises again.

Certainly the Brazilian coast characteristics associated with a closed topography with high solar radiation and temperature of the surface ensures considerable moisture content and low static stability overlying the air, allowing when the synoptic situation is favorable, the occurrence of convective instability and hence surplus moisture in the environment.

The most significant impact will be felt in the Central West and North of Brazil, where there is significant reduction in available moisture over the time slices, especially in the Amazon and Tocantins, Mato Grosso and Goiás, with increased susceptibility to dryness situations.

In quantitative terms (Figure 3), the projection data of the IPCC RCP 8.5 scenario from Eta-Miroc5 show that in the period 1961-1990, over 74% of Brazil had aridity index below 0.9, so moisture supply code. However, modeling shows a reversal of this scenario by 2100, where approximately 67% will be in the next dryness class, that is, between the indexes 0.9 and 2.0. The areas in desertification, north of Bahia and in specific situations of Pará, will suffer a 0.2% increase in area extension until the end of the twenty-first century.

As for the statistical analysis, the Mann-Kendall test indicates a significant trend in the level of 90% of the aridity index for Brazilian regions, except for the Southeast (Table 2). The magnitude of this change will be greater in the North of Brazil, which several studies have pointed to the savannization trend of lush rainforest, which corroborates the results. Jarring the other regions, the South has an increased trend of humidity, that is, over the slices, the trend observed in this region is greater availability of moisture.

Soil moisture data (Figure 4) also confirm the spatial trend in the aridity index, so point to an expansion of the water deficit area, especially in the States of Bahia, Piauí and Minas Gerais. The data also show an available surface moisture reduction for the Midwest Region of Brazil and project a little significant change to the Brazilian coast and the Southeast. However, the State of Mato Grosso do Suland Amazon region have significant loss of soil surface moisture over the time slices, common characteristics by virtue of the types of soils developed in these areas.

These data are of great importance, since they are indicative of water availability to the root system of the plant, which has a direct impact on nutrient absorption capacity of it. Thus, from the modeling data is possible to estimate a reduction of the area with water availability for vegetation development, leading directly to the productivity of various genres that are distributed between the Northeast and Midwest regions of Brazil.

Statistical analyzes show a high significance for the Northeast region of the country, whose magnitude is −0.01% reduction in moisture between the slices of time on study (Table 3). The Region that clashes of this feature is the South of Brazil, whose tendency is soil moisture increases of +0.01% between the slices of time, strong by connecting with the results presented to the aridity index in the same area.

Table 2. Statistical analysis of the aridity index for Brazilian regions. Values in red were considered significant at a confidence level of 90%.

Region	Mann-Kendall	Sen's Bend
North	0.09	+0.04 (by slice)
Northeast	0.09	+0.02 (by slice)
Midwest	0.09	+0.03 (by slice)
Southeast	0.73	+0.01 (by slice)
South	0.09	−0.03 (by slice)

Table 3. Statistical analysis of the soil moisture for Brazilian regions. Values in red were considered significant at a confidence level of 90%.

Region	Mann-Kendall	Sen's Bend
North	0.31	−0.01% (by slice)
Northeast	0.09	−0.01% (by slice)
Midwest	0.31	−0.01% (by slice)
Southeast	0.31	−0.01% (by slice)
South	0.31	+0.01% (by slice)

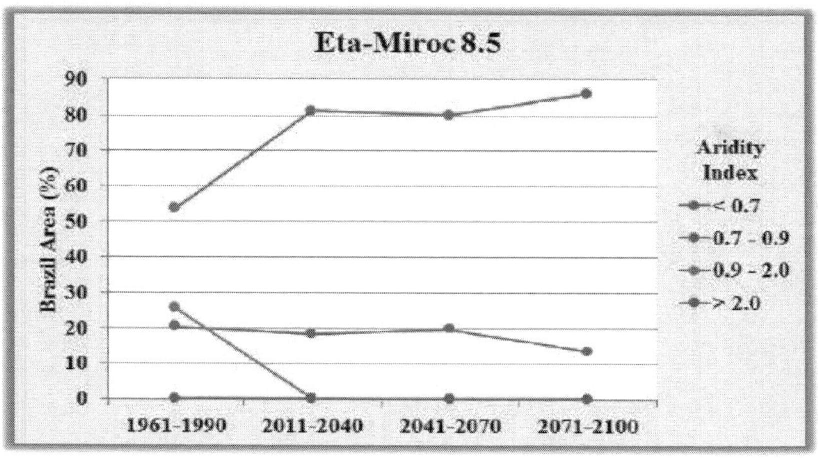

Figure 3. Evolution of the areasofaridity index for Brazil through Eta-Miroc5 regional climate model in the IPCCRCP 8.5 scenario.

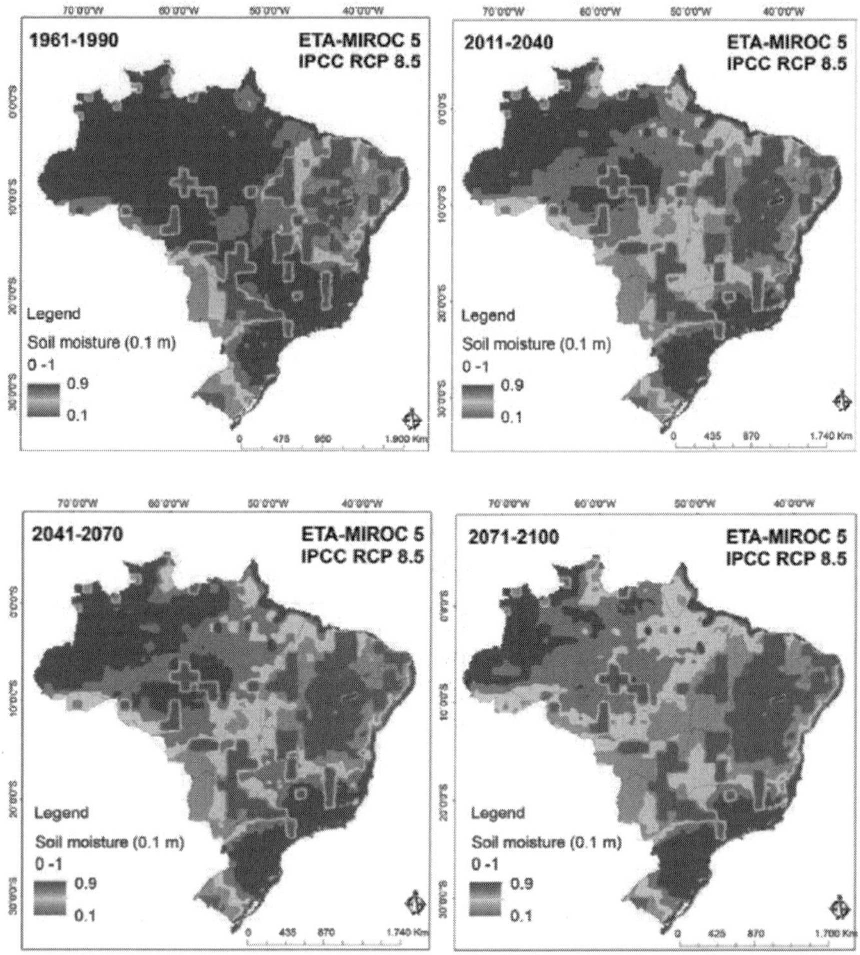

Figure 4. Present climate (1961-1990) and simulated future climate (2011-2040, 2041-2070 and 2071-2100) projected forsoil moisture (0.1 m) in Brazil through Eta-Miroc5 regional climate model in the IPCCRCP 8.5 scenario.

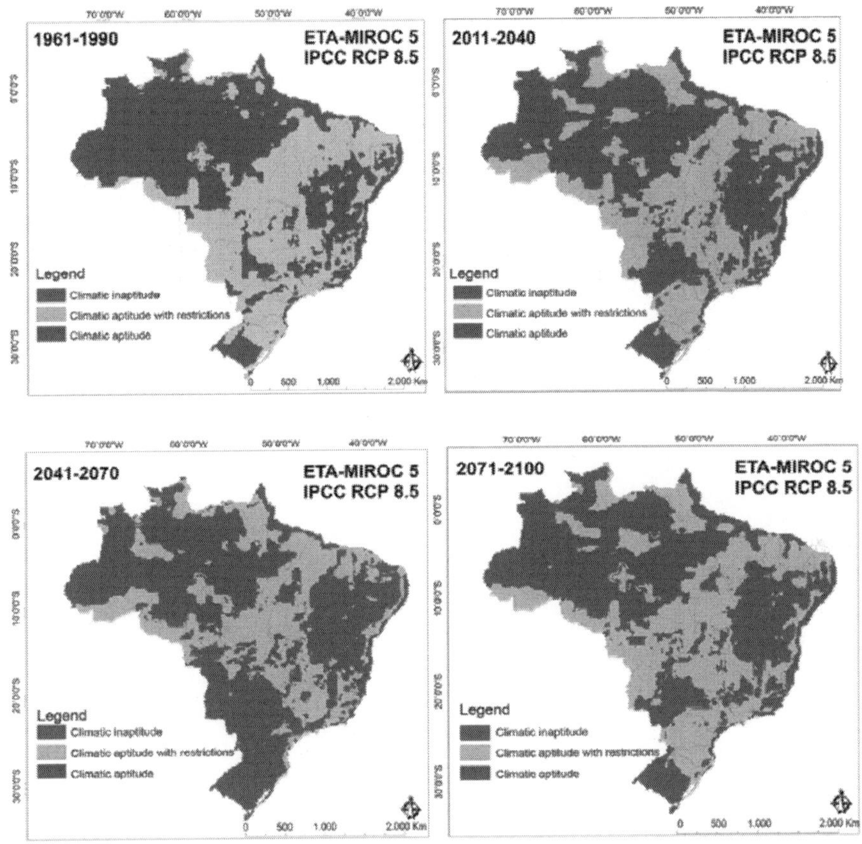

Figure 5. Synthesis map of cultivation fitness class to Brazil for present climate (1961-1990) and simulated future climate (2011-2040, 2041-2070 and 2071-2100) through Eta-Miroc5 regional climate model in the IPCCRCP 8.5 scenario.

From the indicators analyzed above, as well as the air temperature (item 3.1), it was possible the development of Figure 5, which is a synthesis map that shows those areas of Brazil with unfitness for cultivation due to projected climate features (areas in red) as well as those areas which have the potential production, but human interventions with the use of more extensive methods of irrigation are required (yellow areas). The map shows also possible situations of ideal suitability for farming, areas where productivity will reach the maximum possible if the other agronomic and environmental conditions prove favorable (green areas). It is noted that by the end of the century, much of the Northeast of Brazil and in the States of Mato Grosso do Sul and Rio Grande do Sul may present climate unsuited to cultivation. However, most of the coast and the North of the country may be able to farming.

Therefore, from the results presented in this section, it is observed that the Southeast region of Brazil, as well as the country's coastline, is favorable areas for agricultural productivity. However, these regions are historically intended for occupation and development of the urbanization process, that is, with limited land availability and supply for production.

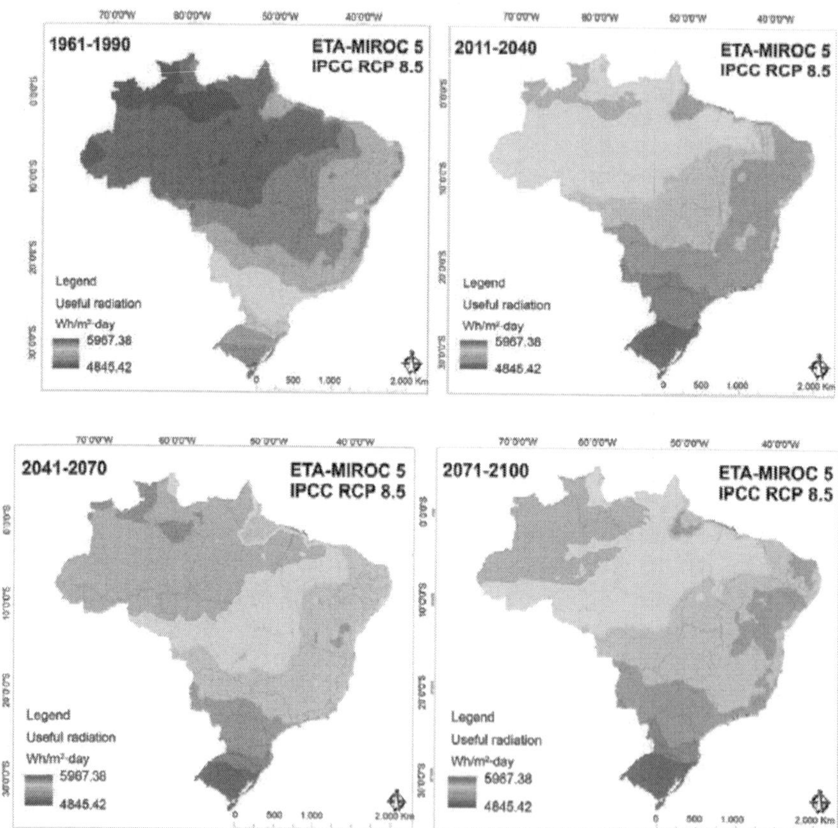

Figure 6. Present climate (1961-1990) and simulated future climate (2011-2040, 2041-2070 and 2071-2100) projected for the useful radiation indicator in Brazil through Eta-Miroc5 regional climate model in the IPCCRCP 8.5 scenario.

3.3. Solar Energy

Analyzing by generalized way all the maps produced by the Eta-Miroc5 regional climate model (Figure 6), there is the same pattern that remains admittedly: an inverse relationship between the intensity of solar radiation and latitude; so at lower latitudes (areas close to the equator, as the North and Northeast of Brazil) solar radiation is high, while in higher latitudes (towards the south of the country), the intensity of this radiation is reduced. This occurs because of the sunlight on the earth's surface, that is, perpendicular in the equatorial and tropical zone and more obliquely toward the poles.

When the historical period (1961-1990) of the Eta-Miroc5 model is evaluated, it is noted that the Northern region of Brazil has the highest values of useful radiation indicator, and these are established between about 5700 and 5900 W·h/m^2·day. Next, there are the Northeast, Midwest and Southeast, with values around 5500 - 5700 W·h/m^2·day. The southern region has the most marked variation of intensity of this radiation, with values ranging between 5500 W·h/m^2·day in the north area to 4900 W·h/m^2·day in southern part.

As mentioned previously, the eminent values of useful radiation indicator in northern Brazil are due to emphatic solar energy that this area receives during the whole year. Thus, despite the yearly averages for the indicator representation, the values of them remain significantly high. In contrast, the southern region exhibits lower values due to the lower amount of solar energy receipted by this area. In addition, the South has the striking feature of quite often the performance of weather systems that enhance cloud cover, especially the cold fronts.

Observing the future projections of useful radiation indicator generated by the Eta-Miroc5 model forced with the IPCC RCP 8.5 scenario, it appears that the simulated values by the end of the twenty-first century are minor compared with the present climate (historical period). On average, such radiation intensity reductions vary between 4% and 8% in Brazil, values considered not significant, since the country still find within a suitable range for the generation of energy from the short-wave radiation (solar radiation).

Between 2011 and 2040, the North and the northern sector of the Midwest Brazil must have an average value of the useful radiation indicator in about 5600 W·h/m^2·day, so a decrease of around 4% compared with the present climate. The south-center of the Midwest Region, as well as the Northeast and Southeast regions are likely to have values around 5300 W·h/m^2·day, representing thus a reduction about 6% over the historical period. The

South should show greater decay in the intensity of solar energy, about 8%, as the average value in this area would be around 5100 W·h/m²·day.

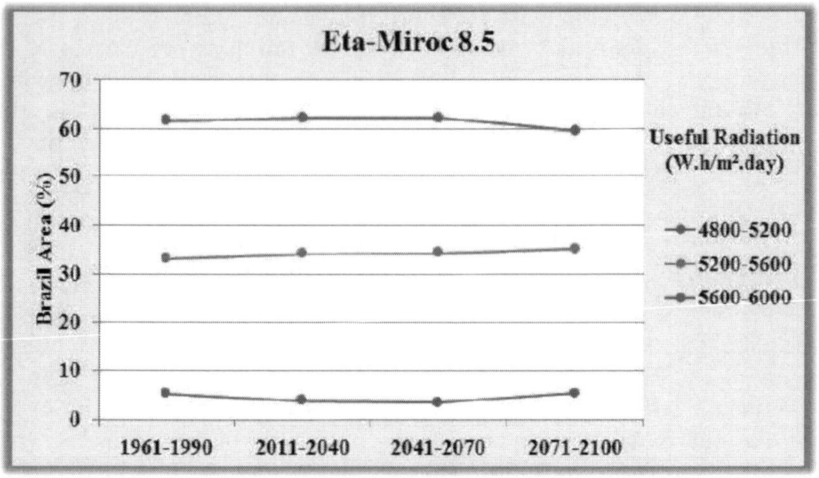

Figure 7. Percentage portions of Brazil area to the useful radiation indicator ranges (W·h/m²·day) between 1961 and 2100 simulated by Eta-Miroc5 regional climate model in the IPCCRCP 8.5 scenario.

Table 4. Statistical analysis of the useful radiation indicator for Brazilian regions. Values in red were considered significant at a confidence level of 90%.

Region	Mann-Kendall	Sen's Bend
North	0.73	−99.56 W·h/m²·day (by slice)
Northeast	0.73	−84.60 W·h/m²·day (by slice)
Midwest	0.73	−92.91 W·h/m²·day (by slice)
Southeast	0.73	−89.57 W·h/m²·day (by slice)
South	0.73	−71.72 W·h/m²·day (by slice)

Accompanying the future projections over 2041 to 2070 periods, there is a negligible increase in the values of useful radiation indicator in Brazil. These elevations are around 2% compared to the previous period (2011-2040), but still remain around 4% lower in relation to the present climate (1961-1990). In the period 2071-2100, the simulated scenario of solar radiation in the country is quite similar to the first discussed future period (2011-2040), with distribution of values of similar manner to that period.

There is only one short exception in the States of Minas Gerais and Paraná, which keeps the value of the previous period (2041-2070).

Figure 7 shows the percentage fractions of Brazil area for certain ranges of useful radiation indicator in function of time simulated by the Eta-Miroc5 regional climate model. Taking into consideration a land area of the country at 8.9 million·km^2, currently (1961-1990) and future (2011-2100) the majority (around 61%) has/will have an average annual radiation intensity between 5600 and 6000 W·h/m^2·day. On a slightly lower power range (5200 - 5600 W·h/m^2·day) it is noted and projected to remain around 34% of the total area of Brazil. Finally, about 5% of the country's land have/will have solar radiation between 4800 and 5200 W·h/m^2·day.

Also according to Figure 6 and Figure 7, it is noted that the range of greater intensity of useful radiation indicator (5600 - 6000 W·h/m^2·day) is located on the North and parts of the Northeast, Midwest and Southeast of Brazil. On the other hand, the range with lower solar radiation (4800 - 5200 W·h/m^2·day) is confined in a given stretch of southern Brazil. This reinforces one of the main geographical features and hence climate of Brazil: a country predominantly tropical with strong solar energy available.

In addition, it is projected a significant increase in average air temperature by the end of XXI century across the country, as seen in item 3.1. It means a reduction in energy conversion capacity, sensitive to ambient temperature variations, especially in the North and Midwest of the country. Furthermore, there would be area limitations for the installation of photovoltaic stations due to the large expansion of the urban phenomenon in the national territory, leaving few free areas for these developments.

In statistical terms, the Mann-Kendal test and Sen's Bend point to a few significant trends to useful radiation indicator in all regions of Brazil (0.73), according to Table 4. Probably, this is due a few factors, including: the use of the 30-year average and the low frequency of variation in solar radiation data. However, from the Sen's Bend analysis, it is noted the magnitude of the trend, with unanimity in whimsy decrease in the useful radiation availability, especially in the North and Midwest, which associated with a rise in temperature could interfere in the production ability of energy from photovoltaic systems. However, it is noteworthy that these values are not statistically significant at the 90% level.

Wherefore, it can be concluded that Brazil has area and highly expressive power for using solar energy, both today and in the coming decades, especially in the North, Northeast and Midwest regions. Regarding these considerations, it is worth noting that, since the given values are annual averages in 30 years, it is expected that in the summer the incident solar radiation intensity is more significant, with attenuation only in the presence of cloudiness, mainly in the North, a factor not impeding in a tropical country.

4. CONCLUSIONS

The objective of this study was to calculate indicators and indexes so that they subsidized the analysis of vulnerability and adaptation of the renewable energy sector to climate change in Brazil, focusing this research specifically in the sources of biofuels and solar energy.

For biofuels, in general, the Brazilian coast is presented as propitious area for agricultural productivity, but these are areas historically intended for occupation and development of the urbanization process, that is, with limited land availability and supply for primary production. In some parts of the Northeast, Midwest and South of the country, offer for the cultivation land will be reduced.

In general, for the solar energy, it was observed that Brazil had area and highly expressive power for the use of this power, both today and in the coming decades, especially in the North, Northeast and Midwest. In statistical terms, the Mann-Kendal test and Sen's Bend point to a very weak tendency to useful radiation indicator in all regions of Brazil. In addition, it is projected a significant increase in mean air temperature by the end of XXI century across the country that can mean a reduction in power conversion capability, which is sensitive to ambient temperature variations, especially in the Midwest and North of the country.

Finally, it should be noted that this work is based on current technology, thus technology improvement with biotechnology incentives can reverse the results once presented, and these are the recommendations of this study: 1) investment in biotechnology and genetic improvement of plants for better adaptation to climatic contrasts; 2) investment in technology to better use of solar energy for power generation, in addition to the improvement of the materials to be less susceptible to temperature variation and transmission systems, which can reduce energy losses in the distribution process.

ACKNOWLEDGEMENTS

The authors acknowledge the support of the Brazilian Ministry of Science, Technology and Innovation within the "Preparation of the Third National Communication of Brazil to the Framework Convention of the United

Nations about Climate Change" project. The authors also thank Rede CLIMA and Sub-Rede Energias Renováveis for the assistance.

CITE THIS PAPER

Antonio OscarJr.,Wanderson LuizSilva,VeraRuffato,RenataBarreto,Marcos Freitas, (2015) Evaluation of Renewable Energy Vulnerability to Climate Change in Brazil: A Case Study of Biofuels and Solar Energy. *Smart Grid and Renewable Energy*,**06**,221-232. doi: 10.4236/sgre.2015.68019

REFERENCES

1. Intergovernmental Panel on Climate Change—IPCC: Climate Change (2013) The Physical Science Basis. Contribution of Working Group I to the Fifth Assessment Report of the IPCC. In: Stocker, T.F., Qin, D., Plattner, G.-K., Tignor, M., Allen, S.K., Boschung, J., Nauels, A., Xia, Y., Bex, V. and Midgley, P.M., Eds., Chapter: Summary for Policymakers, Cambridge University Press, Cambridge, United Kingdom and New York, 1535 p.
2. National Electric Energy Agency—ANEEL (2008) Atlas de Energia Elétrica do Brasil. Agência Nacional de Energia Elétrica, Brazil.
3. Gordon, C., Cooper, C., Senior, C.A., et al. (2000) Simulation of SST, Sea Ice Extents and Ocean Heat Transport in a Version of the Hadley Centre Coupled Model without Flux Adjustments. Climate Dynamics, 16, 147-168.
4. Mesinger, F., Chou, S.C., Gomes, J.L., et al. (2012) An Upgraded Version of the Eta Model. Meteorology and Atmospheric Physics, 116, 63-79.
5. Chou, S., Lyra, A., Mourão, C., et al. (2014) Evaluation of the Eta Simulations Nested in Three Global Climate Models. American Journal of Climate Change, 3, 438-454.
6. Chou, S., Lyra, A., Mourão, C., et al. (2014) Assessment of Climate Change over South America under RCP 4.5 and 8.5 Downscaling Scenarios. American Journal of Climate Change, 3, 512-527.
7. Hare, F.K. (1983) Climate and Desertification: A Revised Analysis. World Climate Applications Programme. World Meteorological Organization, Geneva, 149 p.
8. Klein, S.A. (1977) Calculation of Monthly Average Insolation on Titled Surfaces. Solar Energy, 19, 325-329.
9. Sneyers, R. (1975) Sur L'analyse Statistique des Series Dóbservations. Organisation Méteorologique Mondial, Genève, 192 p.

10. Goossens, C. and Berger, A. (1986) Annual and Seasonal Climatic Variations over the Northern Hemisphere and Europe during the Last Century. Annales Geophysicae, 4, 385-400.
11. Sen, P.K. (1968) Estimates of the Regression Coefficient based on Kendall's Tau. Journal of the American Statistical Association, 63, 1379-1389.

10

CHAPTER

ON THE PERFORMANCE OF A SOLAR DRIVEN ABSORPTION REFRIGERATOR

Ali Fellah , Mouna Hamed, Ammar Ben Brahim

University of Gabes, Engineers National School, Applied Thermodynamic Research Unit, Gabes , Tunisia

ABSTRACT

An endoreversible model is used to simulate the dynamic behavior of a solar driven absorption refrigerator, the cycle under different operating and design conditions. A global time minimization procedure is performed to reach maximum performances. To evaluate the influence of the cold temperature on the system's performances, results are carried out for three values of this temperature. They are presented in normalized charts for general applications. The minimum time set point temperature, entropy and maximum refrigeration load are sharp and therefore, are important to be considered for system design.

Keywords: Refrigeration, Solar Energy, Endoreversibility, Optimization, Transient Regime

1. INTRODUCTION

Refrigeration and air conditioning consume an enormous amount of energy. Solar-powered plants allow the saving of an important amount of energy compared to electric driven ones. Heat driven absorption refrigeration has get more and more attention. Unlike vapor compression refrigerator, absorption refrigerator could be driven by various energy sources. The system uses thermal energy to produce cooling, and thus waste heat and other forms of low grade heat can be employed, especially solar energy. Absolute low noise is also one of their prevailed advantages. Moreover, they cause zero or minimum ozone depletion because they do not use any CFC or HFCs refrigerant as working fluid [1] . From these points of view, the solar-assisted absorption refrigeration shows a very promising future. Usually, the maximum demand for cooling coincides with the maximum availability of solar radiation, whereas conventional electric driven systems have the problem of providing their minimum capacity in the hottest day hours [2] . In addition, the use of solar energy in refrigeration can be very useful to limit the growth of the electric energy demand in summer and for sustaining the development of technologies based on renewable sources of energy. Several authors presented studies of solar refrigeration systems in [3] - [6] .

In the theory of classical thermodynamics, the cyclical model of an absorption refrigerator is treated in [7] as reversible four-heat-sources cycle, which consists of four reversible adiabatic and four reversible isothermal processes. In order to achieve the coefficient of performance (COP) of reversible four-heat-source refrigeration cycles, the isothermal parts of the cycle have to be carried out infinitely slow because of the presence of the thermal resistances. The cooling rate of the resulting refrigerates or trends to zero because an infinite time is taken to draw a finite amount of heat from the cooled space. Real absorption refrigerators are designed to deliver a certain amount of cooling load. It is usually too rough to predict the coefficient of performance of real absorption refrigerators so that it is necessary to develop a new theory for absorption refrigeration cycles. Finite time thermodynamics (FTT) has emerged during the last three decades as a distinct subfield in heat transfer [8] . The method consists of the simultaneous application of heat transfer and thermodynamic principles in the pursuit of realistic models of heat transfer processes, devices, and installations, i.e., models that account for the inherent irreversibility of heat, mass, and fluid flow processes. The original objective of the finite time thermodynamics appears to reduce the differences between the predictions of classical thermodynamics and the results of real practice. The results obtained for varying thermodynamic cycle analyses by using FTT are closer to real device performance than those using classical thermodynamics [9] . In literature, several researches have been focusing on the performance optimization of refrigerators, based on endoreversible models in [10] - [12] and on irreversible models in [13] [14] .

Steady-state models are useful under several conditions, although under strongly dynamic conditions that are often seen in real-life operation, these models are unacceptably and inaccurate [15] . However, steady state models do not provide time dependent information on the thermal behavior of absorption refrigerators and are therefore not suitable for transient system simulations. In order to predict the performance of these chillers under all aspects of operation, dynamic simulations must be developed. Vargas et al. in [16] conducted a simulation of a solar collector driven water heating and absorption cooling plant using the three-heat-reservoir cycles via multi objective optimization. They concluded that the minimum pull-down and pull-up times, and maximum of the second law efficiencies found with respect to the optimized operating parameters are important. Recently, Hamed et al. in [17] presented a thermodynamic transient regime simulation of a solar driven absorption refrigerator. An endoreversible model has been analyzed numerically to find the optimal conditions of a solar driven absorption refrigerator. Based on the latest reference, the main focus of this study is concentrated on the simulation of the transient behavior of the endoreversible model of a three heat reservoirs absorption heat transformer, with Newton's heat transfer law with a detailed solar energy process. For a practical set of cold source temperatures 8°C, 0°C and −8°C, the purpose of this paper is to investigate the effect of solar collector stagnation temperature and cold temperature on time set point temperature, entropy generated inside the cycle and finally evaporator heat.

Solar collectors are broadly classified as, non-concentrating and concentrating type collectors. Flat-plate collectors are the most widely used kind of collectors in the world [18] . In this analysis, the collector adopted is a flat-plate collector. Flat-plate solar collectors are commonly used in solar space heating. It is economically adopted when the same collector is used for both heating and cooling spaces. However, regarding the relatively low temperatures attainable, only a few practical applications are available in flat-plate solar operated cooling processes.

2. MATHEMATICAL MODEL

The solar driven absorption refrigeration cycle and its equivalent model are shown in Figure 1(a) and Figure 1(b). The cycle is considered to be endoreversible, i.e. internally reversible and externally irreversible. Internal irreversibilities are assumed negligible in presence of the heat exchange irreversibilities due to finite temperature differences. Internal irreversibilities due to heat transfer, throttling, mixing and internal dissipation of the working fluid, which are responsible for the entropy generation are always present in a real heat driven refrigerator [4] .

Also, it is assumed that there is no heat loss between the solar collector and the generator and no work exchange occurs between the refrigerator and its environment. The work input required by the solution pump is negligible regarding the energy input to the generator.

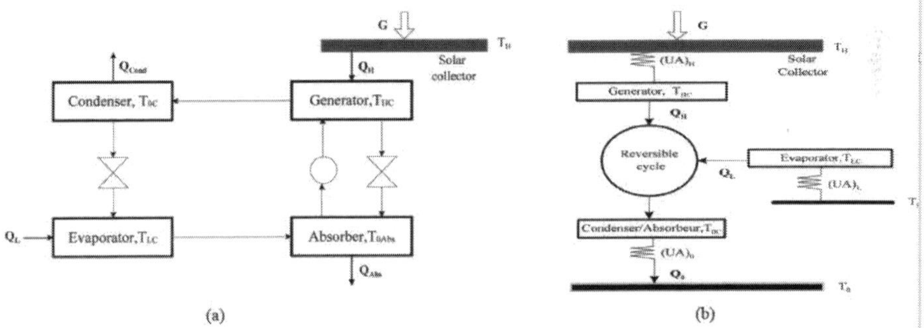

Figure 1. (a) Schematic of a solar driven adsorption refrigeration system; (b) Equivalent endoreversible model.

The heat transfers between the working fluid in the heat exchangers and the external heat reservoirs are carried out under a finite temperature difference and obey the linear heat-transfer law "Newton's heat transfer law". Therefore, the steady-state heat transfer equations for the three heat exchangers can be expressed as:

$$Q_L = (\mathrm{UA})_L (T_L - T_{LC}) \tag{1}$$

$$Q_H = (\mathrm{UA})_H (T_H - T_{HC}) \tag{2}$$

$$Q_0 = (\mathrm{UA})_0 (T_{0C} - T_0) \tag{3}$$

From the first law of thermodynamic:

$$Q_H + Q_L = Q_0 \tag{4}$$

According to the second law of thermodynamic and the endoreversible property of the cycle, we have:

$$\frac{Q_H}{T_{HC}} + \frac{Q_L}{T_{LC}} = \frac{Q_0}{T_{0C}} \tag{5}$$

The generator heat input Q_H could be estimated by the following expression:

$$Q_H = \eta_{sc} A_{sc} G \tag{6}$$

where A_{sc} represents the collector area, G is the irradiance at the collector surface and η_{sc} stands for the collector efficiency. The efficiency of a flat plate collector can be calculated, according to [3] and [5] as:

$$\eta_{sc} = a - b(T_H - T_0) \tag{7}$$

where a and b are two constants.

Equation (7) can be rewritten by introducing the collector stagnation temperature T_{st} as follows:

$$\eta_{sc} = b\left(T_{st} - T_H\right) \tag{8}$$

where T_{st} (for which $\eta_{sc} = 0$) is given by:

$$T_{st} = T_0 + a/b \tag{9}$$

The equation for heat input Q_H can be rewritten by combining Equations (6) and (8) as follows:

$$Q_H = A_{sc} G b\left(T_{st} - T_H\right) \tag{10}$$

The thermal inertia of the refrigerated space is large enough such that the transient operation of the refrigerator can be neglected when compared to the time evolution of the temperature inside the refrigerated space. The transient regime of cooling is accounted for by writing the first law of thermodynamics, as follows:

$$m\, c_{v,\text{air}} \frac{\mathrm{d}T_L}{\mathrm{d}t} = (\text{UA})_w \left(T_0 - T_L\right) + Q_{\text{load}} - Q_L \tag{11}$$

where $(\text{UA})_w (T_0 - T_L)$ is the rate of heat gain from the walls of the refrigerated space and Q_{load} is the load of heat generated inside the refrigerated space.

The factors $(\text{UA})_H$, $(\text{UA})_l$ and $(\text{UA})_0$ represent the overall thermal conductances of the heat exchangers. The overall thermal conductance of the walls of the refrigerated space is given by $(\text{UA})_W$. Since $(\text{UA})_H$, $(\text{UA})_L$ and $(\text{UA})_0$ are not known, the following constraint is introduced at this stage:

$$\mathrm{UA} = (\mathrm{UA})_H + (\mathrm{UA})_L + (\mathrm{UA})_0 \tag{12}$$

According to the cycle model mentioned above, the rate of entropy generated by the cycle is described quantitatively by the second law as:

$$\frac{\mathrm{d}S}{\mathrm{d}t} = \frac{Q_0}{T_0} - \frac{Q_H}{T_H} - \frac{Q_L}{T_L} \tag{13}$$

To provide the model more general, a dimensionless form of the set of equations is preferred, using different reference parameters (as described below). Therefore, it is convenient to search for an alternative formulation that eliminates the physical dimensions of the problem. In a dimensionless model all variables are directly proportional to the dimensional ones presented in Equations (1) to (13).

The complete set of no dimensional equations for the model, including the absorption refrigerator steady state equations is:

$$Q'_L = z(\Gamma_L - \Gamma_{LC}) \tag{14}$$

$$Q'_H = y(\Gamma_H - \Gamma_{HC}) \tag{15}$$

$$Q'_0 = (1 - y - z)(\Gamma_{0C} - 1) \tag{16}$$

$$Q'_H = B(\Gamma_{st} - \Gamma_H) \tag{17}$$

$$Q'_H + Q'_L = Q' \tag{18}$$

$$\frac{Q'_H}{\Gamma_{HC}} + \frac{Q'_L}{\Gamma_{LC}} = \frac{Q'_0}{\Gamma_{0C}} \tag{19}$$

$$\frac{\mathrm{d}\Gamma_L}{\mathrm{d}\theta} = v(1 - \Gamma_L) + Q'_{\mathrm{load}} - Q'_L \tag{20}$$

$$\frac{\mathrm{d}S'}{\mathrm{d}\theta} = Q'_0 - \frac{Q'_H}{\Gamma_H} - \frac{Q'_L}{\Gamma_L} \tag{21}$$

where the following non-dimensional transformations are imposed:

$$\Gamma_H = T_H/T_0, \quad \Gamma_L = T_L/T_0, \quad \Gamma_{st} = T_{st}/T_0,$$
$$\Gamma_{LC} = T_{LC}/T_0, \quad \Gamma_{0C} = T_{0C}/T_0, \quad \Gamma_{HC} = T_{HC}/T_0,$$
$$Q'_H = \frac{Q_H}{\mathrm{UA}T_0}, \quad Q'_L = \frac{Q_L}{\mathrm{UA}T_0}, \quad Q'_0 = \frac{Q_0}{\mathrm{UA}T_0},$$
$$Q'_{\mathrm{load}} = \frac{Q_{\mathrm{load}}}{\mathrm{UA}T_0}, \quad B = \frac{A_{sc} G b}{\mathrm{UA}}, \quad \theta = \frac{t\mathrm{UA}}{m\, c_{\mathrm{v,air}}}. \tag{22}$$

here, B is the parameter which describes the size of the collector relative to the cumulative size of the heat exchangers, and y , z and v are the conductance allocation ratios, defined by:

$$y = \frac{(\mathrm{UA})_H}{\mathrm{UA}}, \quad z = \frac{(\mathrm{UA})_L}{\mathrm{UA}}, \quad v = \frac{(\mathrm{UA})_w}{\mathrm{UA}} \tag{23}$$

According to the constraint property of thermal conductance UA in Equation (12), the thermal conductance distribution ratio for the condenser can be written as,

$$x = \frac{(\mathrm{UA})_0}{\mathrm{UA}} = 1 - y - z \tag{24}$$

3. SIMULATION RESULTS AND ANALYSIS

Using the models presented above, numerical simulations are conducted. The initial conditions are the heat generated inside the refrigerated space $(Q'_{\mathrm{load}} = 0.007)$, conductance allocation and the desired temperature in the refrigerated space. The system behavior is mainly determined by the driven hot temperature and cooling temperature.

The principle main cooling applications are deep freezing $(\Gamma_L = -30 \text{ to } -20^{\circ}\mathrm{C})$, to −20°C), ice making ($\Gamma_L = -5$ to 0°C) and chilling $(\Gamma_L = 5 \text{ to } 15^{\circ}\mathrm{C})$. The refrigerated space temperature to be achieved take three values 8°C, 0°C and −8°C, in dimensionless parameters they are

$\Gamma_{L,\text{set}} = 0.943$, $\Gamma_{L,\text{set}} = 0.916$ and $\Gamma_{L,\text{set}} = 0.889$. For these three cases, the numerical results are shown in Figure 2 to Figure 8. The validated system model can be used to predict system performance at different operating states. The results of computations based on this analysis are various and significant. They point out different operational regimes of the system, namely maximum power output or minimum dissipation (entropy generation) dependently on the studied case. Also, they show the limits of the variation range for the model variables, as well as reasons to choose the best solution between the physical ones.

Time is an important factor in the cooling system for particular cooling period. Hence, at constant heat source temperature Γ_H and for different values of the collector stagnation temperature Γ_{st} the variation of the time set point temperature with the collector size is calculated (Figure 2).

The time set point temperature decreases gradually according to the collector size parameter until reaching a minimum time then it increases in each case of imposed collector stagnation temperature and refrigerated space temperature. The existence of an optimum for the thermal energy input Q'_H is not due to the endoreversible model aspects. However, an optimal thermal energy input Q'_H results when the endoreversible equations are constrained by the recognized total external conductance inventory, UA in Equation (12), which is finite, and the generator operating temperature T_H. These constraints are the physical reasons for the existence of the optimum point. The minimum time to achieve prescribed temperature is the same for different values of the stagnation temperature Γ_{st}. Then, for each imposed value of the stagnation temperature, the dimensionless collector size parameter has a specific variation range decreasing continuously as the collector stagnation temperature increases. For example, at $\Gamma_L = 0.889$ and $\Gamma_{st} = 2.1$, B can vary between 0.337 and 1.126, while at $\Gamma_{st} = 3$, it varies from 0.034 to 0.111.

The results plotted in Figures 3-5 illustrate the minimum time $\theta_{\text{set,min}}$ to reach the desired set point temperature in the refrigerated space and the optimal parameter B_{opt} respectively against the stagnation temperature Γ_{st}, thermal load inside the cold space Q'_{load} and conductance fraction v.

It can be observed from Figure 3 that the minimum time to achieve the prescribed temperature is independent of the collector stagnation temperature. This is due to the heat absorbed by the generator that is quite constant regarding the stagnation temperature variation [17] . If the generator heat input is constant, the dimensionless collector size B in Equation (17) will decrease with higher collector stagnation temperature and the temperature of the generator Equation (15) is also constant. The time set point temperature decreases with the increase of the refrigerated space temperature. For instance, at cold source temperature $\Gamma_{L,set}$ equal to 0.943, 0.916 and 0.889 the set point temperature θ_{set} values are respectively 6.83, 7.85 and 9.45.

This is due to the fact that the temperature of the evaporator starts to decrease linearly then it decreases. Here, the reaction of the evaporator is strongly affected by the generator behavior. Its temperature starts rising linearly, then it becomes stable. As the temperature of the generator is higher more heat is absorbed in the evaporator. While, the temperature of the evaporator decreases slowly, the temperature of the generator is maintained quit constantly, indicating that the equilibrium state has reached. As a result, the optimal dimensionless collector size B decreases monotonically as Γ_{st} increases. Also, one can see that the stagnation temperature Γ_{st} is limited, its value of 1.8 for the first case $(\Gamma_L = 0.943)$, 2.2 for the second case $(\Gamma_L = 0.916)$ and 3 for the third case $(\Gamma_L = 0.889)$ being impossible because of the above mentioned condition regarding the refrigerated space temperature $\Gamma_{L,set}$ desired which is not achieved.

From the curves of Figure 4, as the stagnation temperature Γ_{st} and the heat source temperature Γ_H are constant, there is a clear proportionality between the minimum time set point temperature and the thermal load in the refrigerated space. The minimum time $\theta_{set,min}$ increases respectively when the thermal load Q'_{load} increases and the cold temperature Γ_L decreases. This implies that the minimum time is strongly influenced by the thermal load and the cold temperature. As the load of heat generated inside the cold space is higher, less heat is absorbed at the evaporator, causing a decrease of the temperature of the evaporator. Also the cold temperature decreases with time. Figure 4 reveals that the thermal load inside the cold space has an almost negligible effect on optimal dimensionless collector size, whereas a decrease in the $\Gamma_{L,set}$ leads to a decrease in B_{op}.

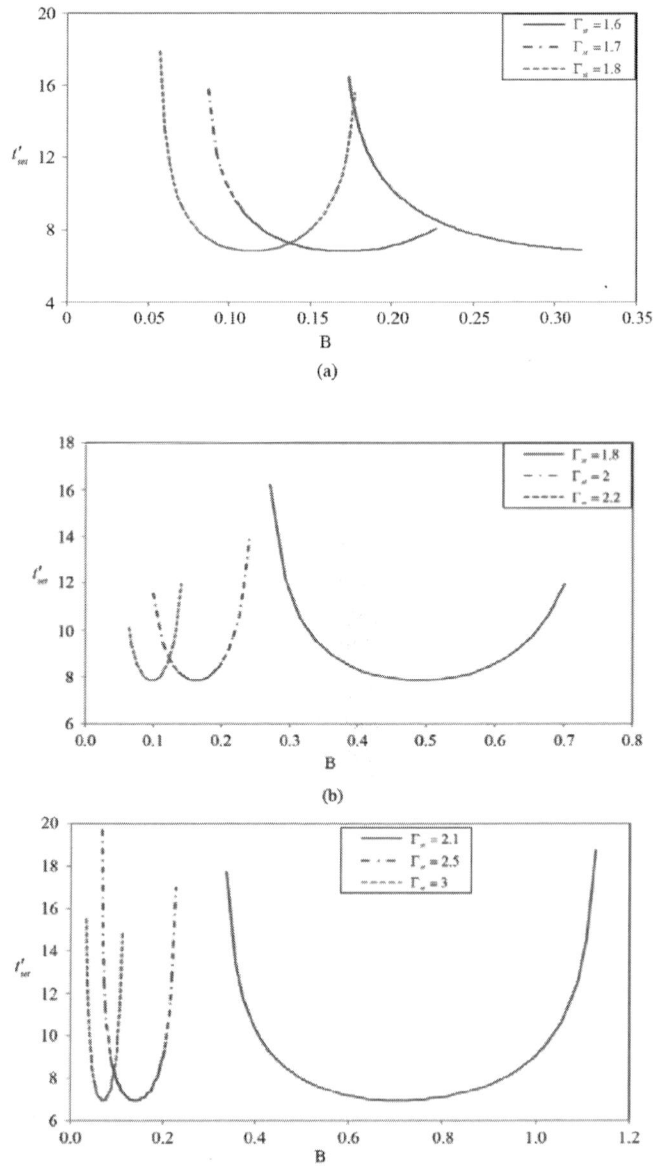

Figure 2. The effect of dimensionless collector size B on time set point temperature. (a) $\Gamma_{L,set} = 0.943$, $\Gamma_H = 1.5$; (b) $\Gamma_{L,set} = 0.916$, $\Gamma_H = 1.5$; (c) $\Gamma_{L,set} = 0.889$, $\Gamma_H = 1.5$.

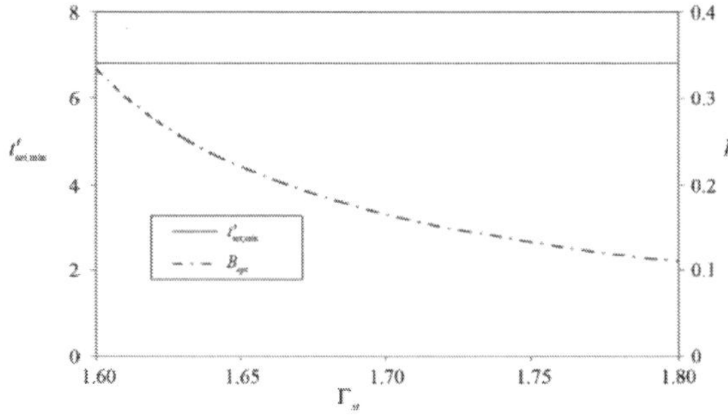

(a)

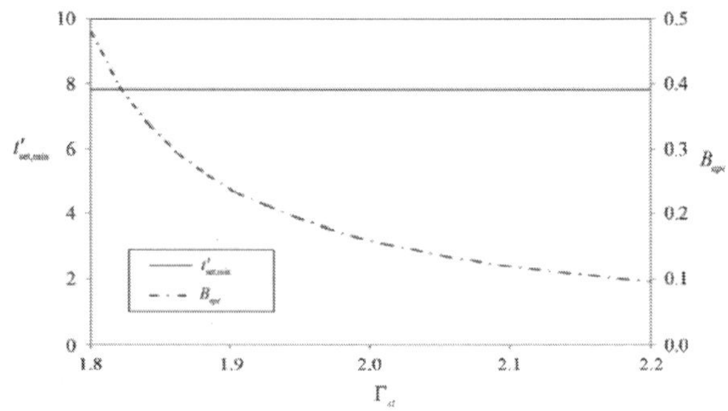

(b)

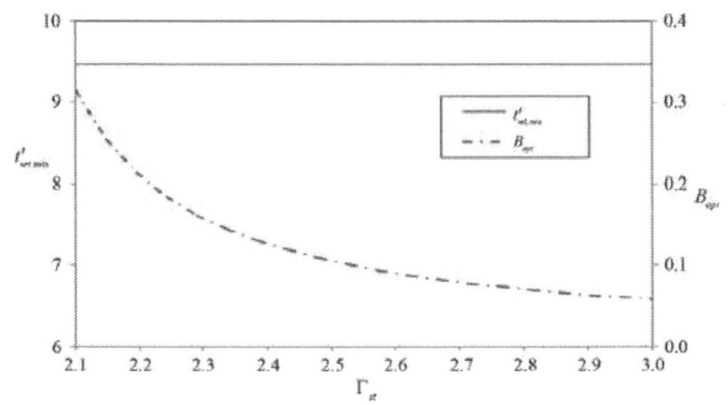

(c)

Figure 3. The effect of collector stagnation temperature on minimum time set point temperature and optimal collector size. (a) $\Gamma_{L,set} = 0.943$, $\Gamma_H = 1.5$; (b) $\Gamma_{L,set} = 0.916$, $\Gamma_H = 1.7$; (c) $\Gamma_{L,set} = 0.889$, $\Gamma_H = 2$.

Similar characteristics were obtained concerning the behaviors of the set point temperature and the optimal collector size according to conductance allocation ratios, as shown in Figure 5.

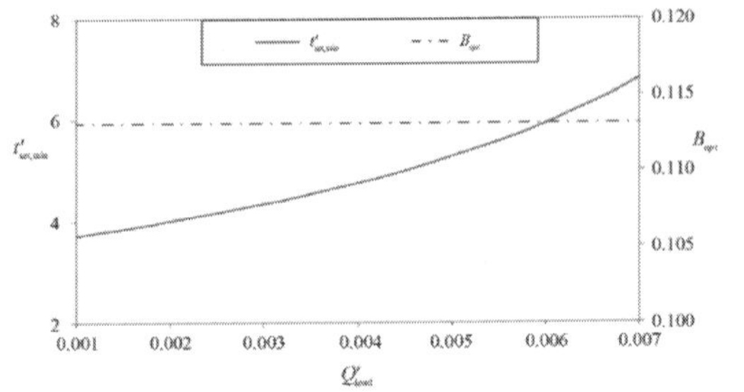

(a)

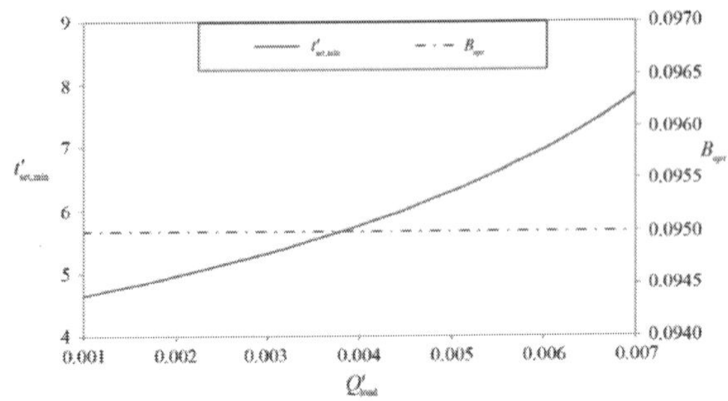

(b)

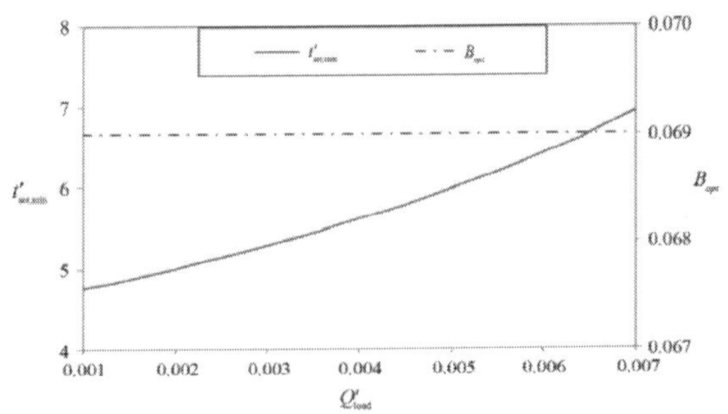

(c)

Figure 4. The effect of thermal load in the refrigerated space on minimum time set point temperature and optimal collector size.
(a) $\Gamma_{L,\mathrm{set}} = 0.943$, $\Gamma_H = 1.5$, $\Gamma_{st} = 1.5$; (b) $\Gamma_{L,\mathrm{set}} = 0.916$, $\Gamma_H = 1.5$, $\Gamma_{st} = 1.8$; (c) $\Gamma_{L,\mathrm{set}} = 0.889$, $\Gamma_H = 1.5$, $\Gamma_{st} = 2$.

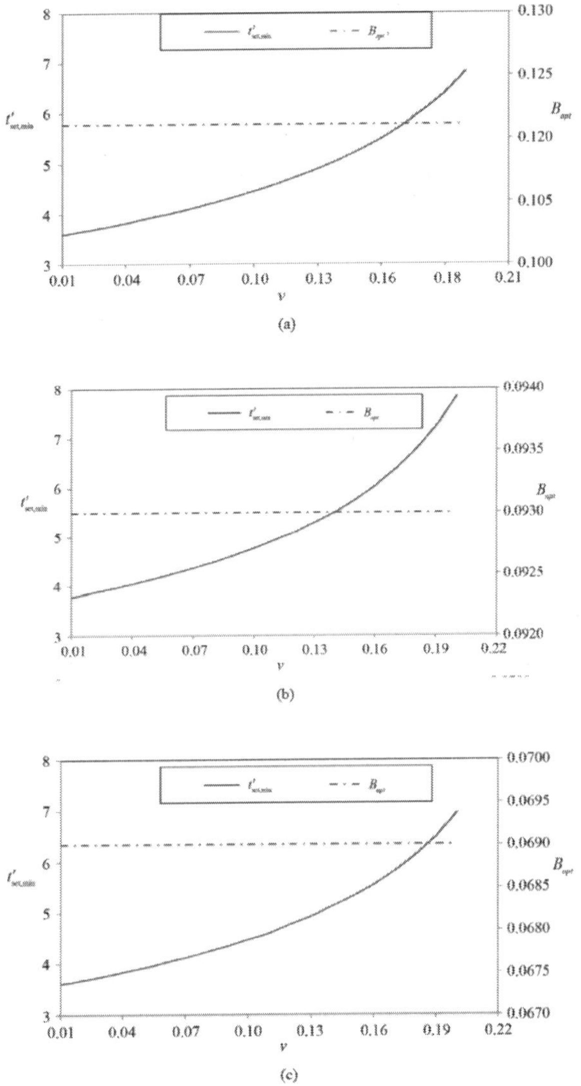

Figure 5. The effect of conductance fraction on minimum time set point temperature and optimal collector size.
(a) $\Gamma_{L,\mathrm{set}} = 0.943$, $\Gamma_H = 1.5$, $\Gamma_{st} = 1.8$; (b) $\Gamma_{L,\mathrm{set}} = 0.916$, $\Gamma_H = 1.5$, $\Gamma_{st} = 1.8$; (c) $\Gamma_{L,\mathrm{set}} = 0.889$, $\Gamma_H = 2$, $\Gamma_{st} = 3$.

The total entropy generated inside the cycle function up to θ_{set} given in Equation (21) can be plotted with respect to the dimensionless collector size B for various stagnation temperature parameters Γ_{st} and cold temperature and at constant heat source temperature as shown in Figure 6.

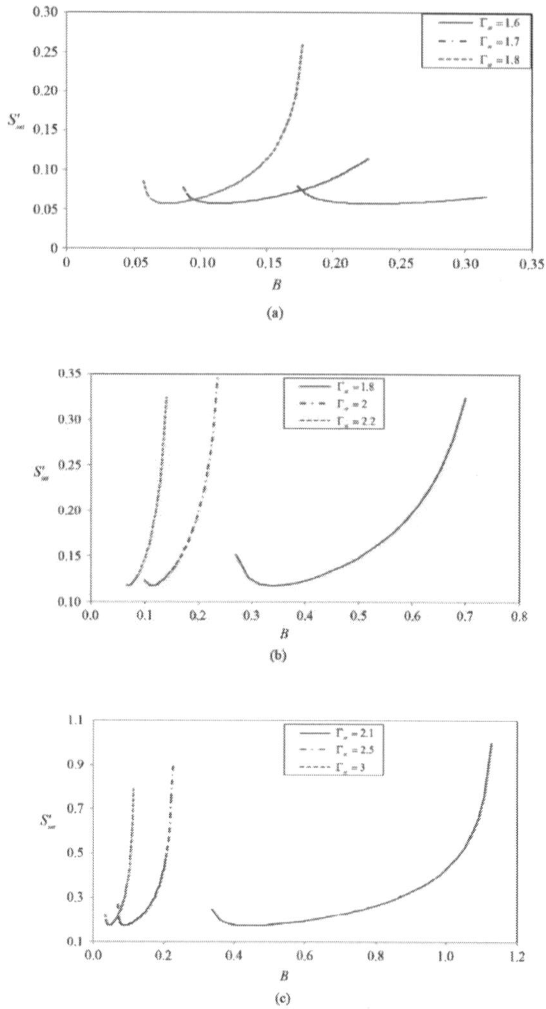

Figure 6. Total entropygenerated to reach a refrigerated space set point temperature.
(a) $\Gamma_{L,set} = 0.943$, $\Gamma_H = 1.5$; (b) $\Gamma_{L,set} = 0.916$, $\Gamma_H = 1.7$; (c) $\Gamma_{L,set} = 0.889$, $\Gamma_H = 2$.

A specific B value exists that minimizes the entropy function for given Γ_{st} and Γ_L values. The curves show clearly a reverse proportionality between the internal entropy generation and the stagnation temperature. Then, for each imposed value of the stagnation temperature, the dimensionless collector size has a specific variation range (imposed by the condition S > 0) continuously decreases as the efficiency increases.

The optimization with respect to the stagnation temperature is pursued in Figure 7 for minimum total entropy generated up to set point temperature and optimal dimensionless collector size. It is observed that $S'_{set,min}$ is independent of Γ_{st}, whereas a decrease in the cold temperature leads to an increase in $S'_{set,min}$. For instance, at a cold source temperature $\Gamma_{L,set}$ equal to 0.943, 0.916 and 0.889 the minimum total entropy generated values are respectively 0.018, 0.118 and 0.229. The cold temperature decreases with time. The total entropy inside the cycle

is strongly affected by the time. It rises with the increase of time. On the basis of the second law of thermodynamics, the entropy production is always positive for an externally irreversible cycle.

From Figure 7, when the stagnation temperature increases then the collector optimal size decreases. This result brings to light the need for delivering towards the greatest values of Γ_{st} to approach the real refrigerator. However, the stagnation temperature Γ_{st} is limited.

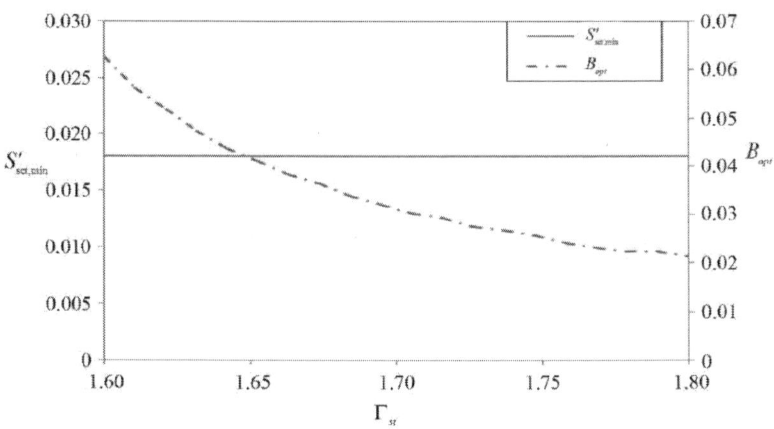

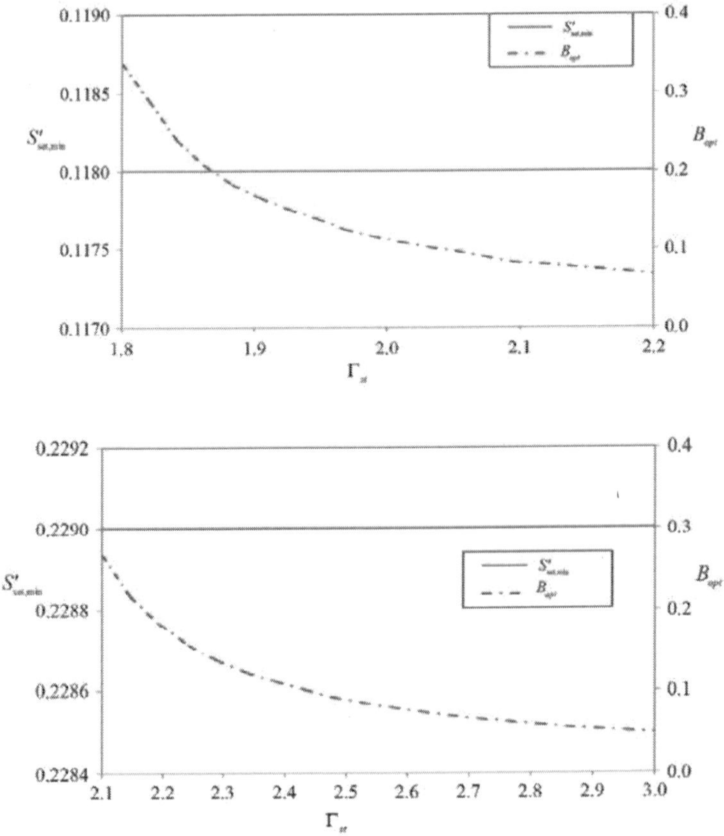

(c)

Figure 7. The effect ofdimensionless collector stagnation temperature on minimum entropy set point temperature and optimal collector size

(a) $\Gamma_{L,set} = 0.943$, $\Gamma_H = 1.5$; (b) $\Gamma_{L,set} = 0.916$, $\Gamma_H = 1.5$; (c) $\Gamma_{L,set} = 0.889$, $\Gamma_H = 2$.

In this final section, the solar absorption refrigerator is considered to operate according to a closed cycle at steady state. Figure 8 reveals clearly the relationship between the evaporator heat with dimensionless collector size at constant heat source temperature and for three different values respectively of the collector stagnation temperature and the cold source temperature. It can be identified from the results that the cooling capacity increases from the startup of the refrigeration system. This phenomenon is typical of the system start-upworking.

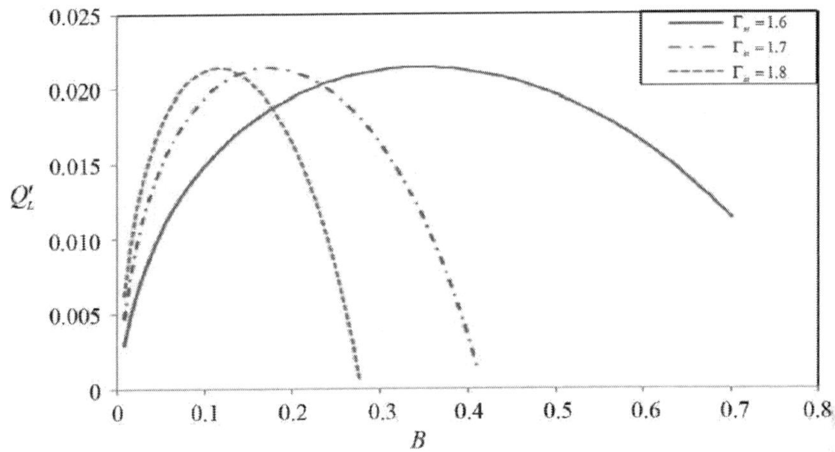

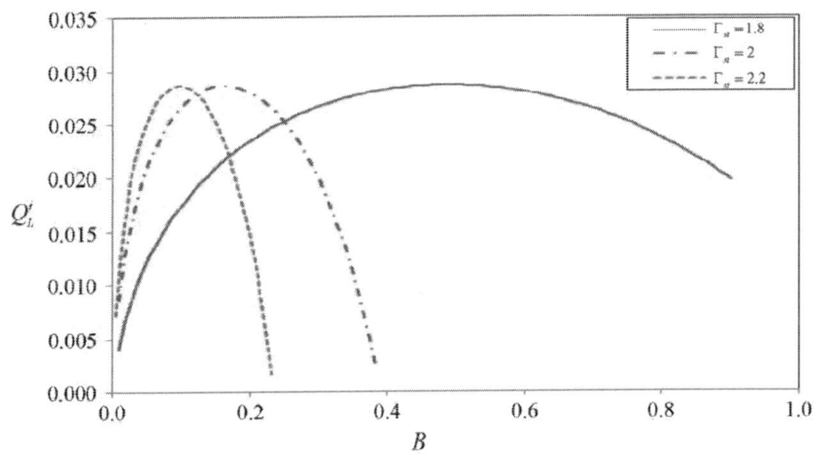

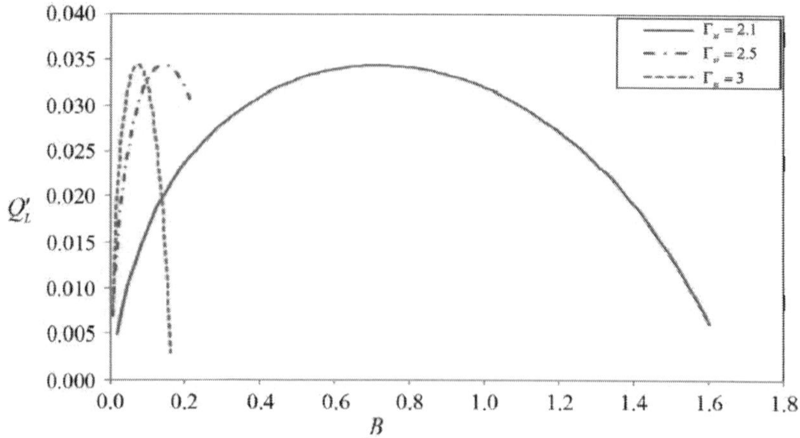

(c)

Figure 8. The effect of dimensionless collector size B on the heat absorbed by the evaporator.
(a) $\Gamma_{L,\text{set}} = 0.943$, $\Gamma_H = 1.5$; (b) $\Gamma_{L,\text{set}} = 0.916$, $\Gamma_H = 1.8$; (c) $\Gamma_{L,\text{set}} = 0.889$, $\Gamma_H = 3$.

regime when there is a refrigerant hot flow inside the evaporator that would heat the interior while the refrigeration system cooling capacity increases. When the steady state conditions are attained in the refrigerator the evaporator heat transfer reaches a maximum. The maximum cooling capacity obtained for the tested refrigerator remains constant with the stagnation temperature. Then, for each imposed value of Γ_{st}, Q_L, has a specific variation range continuously decreasing while B increases. The maximum evaporator heat transfer variation shown by the curves increases when the cold temperature decreases, as is demonstrated with the result

for $\Gamma_{L,\text{set}} = 0.943$, $Q_{L,\max} = 0.0214$, $\Gamma_{L,\text{set}} = 0.916$, $Q_{L,\max} = 0.0287$, and

finally $\Gamma_{L,\text{set}} = 0.889$, $Q_{L,\max} = 0.0344$, . It may be concluded that $Q_{L,\max}$ is more affected by $\Gamma_{L,\text{set}}$ than by Γ_{st}.

4. CONCLUSIONS

In this paper a dynamic model of a solar driven absorption refrigerator has been presented. An externally irreversible but internally reversible model has been employed to analyze the optimum conditions for which the maximum refrigeration effect can be achieved.

The cycle analysis has been studied under standard and variable working conditions. The results also supplied the initial conditions of the simulation of absorption performances. The existence of a collector optimal size for minimum time to reach a specified temperature in the refrigerated space, minimum entropy generation inside the cycle and maximum refrigeration rate was demonstrated. Appropriate dimensionless groups were identified and the generalized results reported in dimensionless charts.

The following conclusions could be derived:

- The minimum time to reach the prescribed temperature in the refrigerated space depends strongly on the cold temperature and the thermal load on the cold space but remains constant with the stagnation temperature variation.
- The minimum time set point temperature and collector optimal dimensionless size are influenced analogously by the thermal load in the refrigerated space and by the thermal conductance of the walls.
- $S'_{set,min}$ is independent of Γ_{st}, whereas a decrease in $\Gamma_{L,set}$ leads to an increase in $S'_{set,min}$.
- The maximum evaporator heat transfer increases when the cold temperature decreases, but remains constant as the stagnation temperature increases.

CITE THIS PAPER

AliFellah,MounaHamed,Ammar BenBrahim, (2014) On the Performance of a Solar Driven Absorption Refrigerator. *Energy and Power Engineering*, **06**,278-291. doi: 10.4236/epe.2014.69024

REFERENCES

1. Abdullah, M.O. and Hien, T.C. (2011) Comparative Analysis of Performance and Techno-Economics for a H2O-NH3-H2 Absorption Refrigerator Driven by Different Energy Sources. Applied Energy, 87, 1535-1545.
2. Calise, F., Denticed'Accadia, M. and Vanoli, L. (2011) Thermoeconomic Optimization of Solar Heating and Cooling Systems. Energy Conversion and Management, 52, 1562-1573.
3. Bejan, A. and Vargas, J.V.C. (1995) Optimal Allocation of a Heat Exchanger Inventory in Heat Driven Refrigerators. International Journal of Heat and Mass Transfer, 38, 2997-3004.
4. Chen, J. and Schouten, J.A. (1998) Optimum Performance Characteristics of an Irreversible Absorption Refrigeration System. Energy Conversion and Management, 39, 999-1007.
5. Sokolov, M. and Hershgal, D. (1993) Solar-Powered Compression-Enhanced Ejector Air Conditioner. Solar Energy, 51, 183-194.
6. Vargas, J.V.C., Sokolov, M. and Bejan, A. (1996) Thermodynamic Optimization of Solar-Driven Refrigerators. Journal of Solar Energy Engineering, 118, 130-135.
7. Huang, F.F. (1976) Engineering Thermodynamics. Macmillan, New York.
8. Bejan, A. (1995) Theory of Heat Transfer-Irreversible Power Plants-II. The Optimal Allocation of Heat Exchange Equipment. Journal of Heat and Mass Transfer, 38, 433-444.
9. Denton, J.C. (2002) Thermal Cycles in Classical Thermodynamics and Non Equilibrium Thermodynamics in Contrast with Finite Time Thermodynamics. Energy Conversion and Management, 43, 1583-1617.
10. Fellah, A., Khir, T. and Brahim, A.B. (2010) Hierarchical Decomposition and Optimization of Thermal Transformer Performances. Structural and Multidisciplinary Optimization, 42, 437-448.
11. Bautista, O., Mendez, F. and Cervantes, J.G. (2003) An Endoreversible Three Heat Source Refrigerator with Finite Heat Capacities. Energy Conversion and Management, 44, 1433-1449.

12. Wijeysundera, N.E. (1999) Simplified Models for Solar-Powered Absorption Cooling Systems. Renewable Energy, 16, 679-684.
13. Goktun, S. (1997) Optimal Performance of an Irreversible Refrigerator with Three Heat Sources. Energy, 22, 27-31.
14. Chen, J. and Wu, C. (1996) The R-? Characteristics of a Three-Heat-Source Refrigeration Cycle. Applied Thermal Engineering, 16, 901-905.
15. Brown, M.W. and Bansal, P.K. (2002) Transient Simulation of Vapour-Compression Packaged Liquid Chillers. International Journal of Refrigeration, 25, 597-610.
16. Vargas, J.V.C., Ordonez, J.C., Dilay, A. and Parise, J.A.R. (2009) Modeling, Simulation and Optimization of a Solar Collector Driven Water Heating and Absorption Cooling Plant. Solar Energy, 83, 1232-1244.
17. Hamed, M., Fellah, A. and Ben Brahim, A. (2012) Optimization of a Solar Driven Absorption Refrigerator in the Transient Regime. Applied Energy, 92, 714-724.
18. Chidambaram, L.A., Ramana, A.S., Kamaraj, G. and Velraj, R. (2011) Review of Solar Cooling Methods and Thermal Storage Options. Renewable and Sustainable Energy Reviews, 15, 3220-3228.

11

CHAPTER

THE APPLICATION STUDY IN SOLAR ENERGY TECHNOLOGY FOR HIGHWAY SERVICE AREA: A CASE STUDY OF WEST LUSHAN HIGHWAY LOW-CARBON SERVICE AREA IN CHINA

Xiaochun Qin,[1,2] Yi Shen,[1] and Shegang Shao[1]

[1]*Research Institute of Highway, Ministry of Transport, Beijing 100088, China*
[2]*Tsinghua University, Beijing 10084, China*

ABSTRACT

A lot of research works have been made concerning highway service area or solar technology and acquired great achievements. However, unfortunately, few works have been made combining the two topics together of highway service areas and solar energy saving to make a systemic research on solar technology application for highway service area. In this paper, taking West Lushan highway low-carbon service area in Jiangxi Province of China as the case study, the advantages, technical principles, and application methods of solar energy technology for highway service area including solar photoelectric technology and solar water heating technology were discussed based on the analysis of characteristics of highway low-carbon service area; the system types, operation mode, and installing tilt angle of the two kinds of solar systems suitable for highway service areas were confirmed. It was proved that the reduction of the cost by electricity savings of solar system was huge. Taking the investment of the solar systems into account, the payback period of solar photoelectric systems and solar water heating systems was calculated. The economic effect of the solar systems in West Lushan highway service area during the effective operation periods was also calculated and proved very considerable.

1. INTRODUCTION

As the renewable and clean energy, solar energy will not cause environmental pollution that is produced by traditional energy like coal, oil, and other fossil fuels during the using process. Development and utilization of green energy are one of the most important energy saving measures. During the past 30 years, many developed industrial countries and some developing countries have attached great importance to the development of solar technology and solar technology has been widely promoted and used in residential areas. The pace of research and application of solar energy in Europe is also very fast and the use of solar heat is more extensive, involving power generation, domestic hot water, heating, local heating water, and so forth. More than 90% of the solar heat is used for residential areas in the EU [1].

China began to use solar energy early in the 1970s, achieving good results in the beacon lights. In the 1990s, solar photovoltaic technology began to be used in road lighting and signals and solar water heating system entered the rural residential areas under the support of China government by the late 20th century. In the 21st century, solar photovoltaic technology has been greatly promoted by the growing attention of government and the improvement of solar cell production technology, meanwhile solar energy applications field also expanded gradually.

Nowadays, solar technology is gradually extended to highway traffic field with the introduction of new traffic lighting devices like solar traffic lights, solar orientation lights, solar street lamps, and so forth. However, few solar energy technologies have been used in highway service area in China. As the basic infrastructure serving for vehicles and passengers, the overall function of highway service area determines the quality and efficiency and the economic benefits of the service. Because of the far location away from urban areas, generally highway service areas have few external energy sources to use, while also consuming more energy. It is relatively closed and independent system. The closeness and independence of highway service area determines its dependence and pressure on resources and environment. The application of solar technology in the construction of highway service area could not only alleviate the scarcity of resources, reduce energy consumption, and improve efficiency of resource and energy, but also reduce environmental pollution, maintain ecological balance, improve highway operational efficiency, and optimize the service quality.

The extensive researches on highway service area began aboard in the early 1990s. American Association of State Highway and Transportation (AASHTO) published the third edition of "expressway service construction guide" in 2001, which stressed the application of green building technology from the aspects of domestic sewage, garbage, and so forth. The studies in the field of highway service facilities in Japan defined the planning layout, architectural forms, design principles, and so on comprehensively, clearly, and in detail. According to "Japan Highway Design Manual" of 1980 version, the design essentials of service rest facilities were defined. The technical standards and design methods of highway new rest facilities are made in "Japanese highway design essentials" in 1991 from more comprehensive aspects. The researches on highway service area were launched relatively late in China. The function, size, and technical requirements of service facilities were simply regulated in "Technical Standards of Highway Construction (JTG B01-2003)" [2], without involving the energy saving technology. Hui and Liu [3] analyzed the space size and parking capacity of highway rest facilities according to the field survey data and theoretical calculations for Shenda Highway service area in Northeast China. Luo [4] made a study in the function, location, spacing control, and architectural design requirements of highway service. Wu [5] proposed the biological disposal method of highway service area sewerage for vegetation watering and toilet flushing. Yang [6] discussed the economic effects of biological contact oxidation treatment technology using highway service area sewage disposal. It could be seen that the researches on highway service area in China focused mainly on the layout and scale of the facilities, the operation way, water supply and sewage treatment, and so forth but few on energy saving technology.

In the meanwhile, many researches were made on the economic effect assessment of solar technology. Maria and Marano [7] discussed the economic analysis methods of solar photovoltaic systems under different climatic conditions and pointed out that the economic analysis methods were related to local policy as well as technical level and climatic conditions. Schröder and Reddemann [8] made an analysis in the economic influence of hot water consumption and energy efficiency on solar collectors in different months and climatic conditions in terms of the German federal government's economic conditions. Zilla and Abraham [9] proved that the economic effects became better with the increase of useful life period, fuel prices, and market demand based on the establishment of a model for the economic practicability of solar hot water systems and also concluded that the effect of series solar collectors is better than that of parallel collectors. Hawlader et al. [10] made the economic evaluation for solar heating water system by using different variables and concluded that the designed collector area is 1000 m^2 for best payback period and internal rate of return. Yi [11] made an analysis in the energy consumption constitution and energy saving way for residential building. It can be seen that solar technology brings very considerable economic benefits as well as reducing environmental pollution.

In summary, a lot of research works have been made related to the two fields of highway service area and solar technology and acquired great achievement. However, unfortunately, few works were made combining the two topics together of highway services and solar energy saving to make a systemic research on solar technology in green highway service area due to the design and construction of the service mainly undertaken by the transport authorities, coupled with the faraway location of the service and its subordinate position in the highway system. In this context, it is quite necessary and urgent for launching the research on solar technology and economic effects of highway service area to promote low-carbon highway service area development and guide solar engineering practice.

2. OBJECTIVES

The main objectives of this paper were as follows:(1)confirming the applicability and advantages of solar technology for highway service;(2)choosing the suitable type and the best tilt angle of solar photovoltaic system for highway service area based on the location and local climate conditions;(3)choosing the suitable operation type of solar water heating system, suitable type, and the best installing angle of collectors for highway service area;(4)calculating the economic effects and environmental effects of solar technologies on highway service area.

3. METHODS

3.1. The General Situation of Low-Carbon Highway Service Area

In recent years, with the development of green transportation ideas in China the concept of low-carbon highway service area has been proposed and people are increasingly concerned about the comfort and green level of the service area. Low-carbon highway services generally refer to the services which integrate various measures throughout the entire life cycle of planning, designing, construction, operation, and management to maximize saving resources, protect environment, reduce pollution, and provide people with a safe, healthy, comfortable, and efficient highway service area based on the principle of the virtuous circle of ecosystem in design conception and construction practice. Some highway service areas have begun to employ different energy saving technologies to try establishing low-carbon service areas in China. Green building and clean energy technologies have been used in a small number of service areas and achieved good effects [12].

In this paper, utilization technologies and methods of solar photoelectric and water heating technology for highway service area were analyzed according to the natural geographical and climatic conditions of West Lushan scenic area, which would achieve effective operational cost reductions and good energy saving effects.

West Lushan highway is located in northern Lushan Scenic Area in Jiangxi Province of China, with the extremely rich and natural landscape resources along and it is the highway with the highest environmental protection requirements in Jiangxi as seen in Figure 1. The location of West Lushan highway service area is in the southern region and has rich solar energy resources with 4494.35 MJ/m^2 (horizontal) annual radiation and 1700.7 sunshine hours. In the construction of West Lushan service area, solar energy technologies are fully made in use in order to maximize energy savings and environment protection.

Figure 1: The aerial view of Lushan highway service area.

3.2. Solar Photoelectric Technology for Highway Service Area

3.2.1. Technical Principles and Constitution

Solar photoelectric technology is based on the photoelectric effect, through the using of solar cells to convert solar energy directly into electricity, as shown in Figure 2. There are two main types of solar photovoltaic systems: independent photovoltaic power generation and grid-connected photovoltaic power generation.

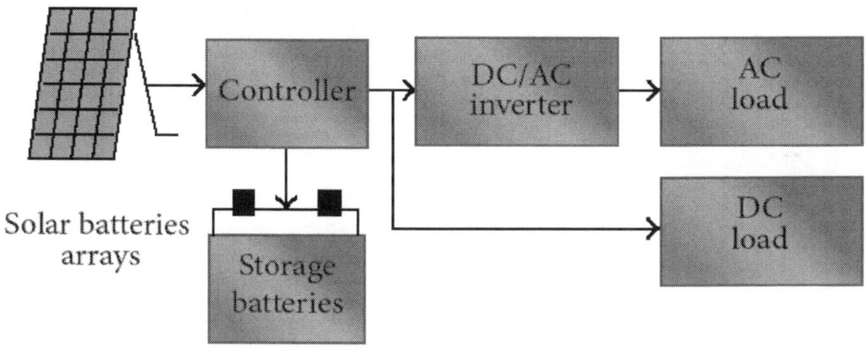

Figure 2: The diagrammatic sketch of solar photoelectric technology.

Photovoltaic power generation system consists of three major components of solar panels, controller, and inverter. Solar panel is the core part of the system and its role is to convert solar radiation to electrical energy stored in batteries or generated. Solar controller is to control the working status of the entire system and make an overcharge or overdischarge protection for battery [13]. The role of the inverter is to convert DC (direct current) solar power into AC (alternating current) power, directly transmitting 12 VDC, 24 VDC, and 48 VDC solar energy to provide 220 VAC electrical power [14].

3.2.2. Advantages of Solar Photovoltaic (PV) System for Highway Service Area

They are as follows (1) small size, light weight, and small occupational space like building integrated photovoltaic (BIPV);(2)long useful life period: the life of PV systems can reach 20 to 50 years;(3)zero pollution emissions:

PV consumes no fuel with no noise and no pollution to the service area;(4)power generation without water: PV systems can be applied for highway service area in the uninhabited areas.

3.2.3. The Selection of Solar Photovoltaic System Type for Highway Service Area

Highway service area is usually to use independent photovoltaic systems, with the appropriate choice of voltage levels and the number of phases in accordance with electricity requirements, as shown in Table 1 [15].

Table 1: Design and selection of solar photovoltaic system.

System type	Type of current	Whether it has the energy storage device	Application scope
Stand-alone photovoltaic system	Direct current (DC)	Yes	Remote areas with no power, DC power load equipment, and high power continuity requirements for power supply
		No	Remote areas with no power, DC power load equipment, and no continuity requirements for power supply
	Alternating current (AC)	Yes	Remote areas with no power, AC power load equipment, and high power continuity requirements for power supply
		No	Remote areas with no power, AC power load equipment, and no continuity requirements for power supply

3.2.4. The Determination of the Tilt Angle of Solar Photovoltaic Systems for Highway Service Area

In the photovoltaic system, the solar radiation received by PV arrays differs according to its tilt angle. For a fixed PV array, if the designed inclined plane is not rational, there may be a waste of more solar modules to meet the power or causes of the unreasonable capacity of battery, which makes the cost of the entire system higher. For determining the best angle of PV arrays, the nature of the load, local weather, and geographical conditions should be taken into account. There are three kinds of load of PV systems including balanced, seasonal, and temporary load. Balanced load refers to the daily electricity for production and life. Air conditioning is the seasonal load and temporary load is generally for the unexpected emergency electrical events. In general, the tilt angle of PV system is determined by application amount of the balanced load and the specific load condition should also be considered. The basic principle is to reach the best tilt angle, reduce the instability of the solar radiation throughout the year, and make the solar radiation amount more evenly for the entire year received by inclined planes.

Research [15] has showed that the designed installing tilt angle of photovoltaic panels equals the local latitude ± 15°. Considering the average amount of electricity use throughout the year, the tilt angle = local latitude; considering the amount of electricity use in summer months (April to September), the tilt angle = local latitude − 15°; considering the amount of electricity use in winter months (October to March), the tilt angle = local latitude + 15°. And the specific azimuth and tilt angle should be considered together for the actual situation of the buildings in service areas.

3.3. Solar Water Heating Technology for Highway Service Area

3.3.1. Technical Principles and Characteristics

Solar heating water system is the system converting solar radiation into thermal energy to heat water and delivering to each user and typically includes a solar collector, heat storage, circulation pumps, pipe connections, brackets, and other components. The system constitutional diagram is shown in Figure 3.

There are three types of solar water heating system including natural circulation systems, DC systems, and forced circulation system. According to the presence or absence of heat exchanger, solar water heating systems are divided into two types of direct and indirect systems. According to hot water providing range, solar water heating systems are divided into three types of centralized heating water systems, centralized-decentralized heating water systems and decentralized water heating systems.

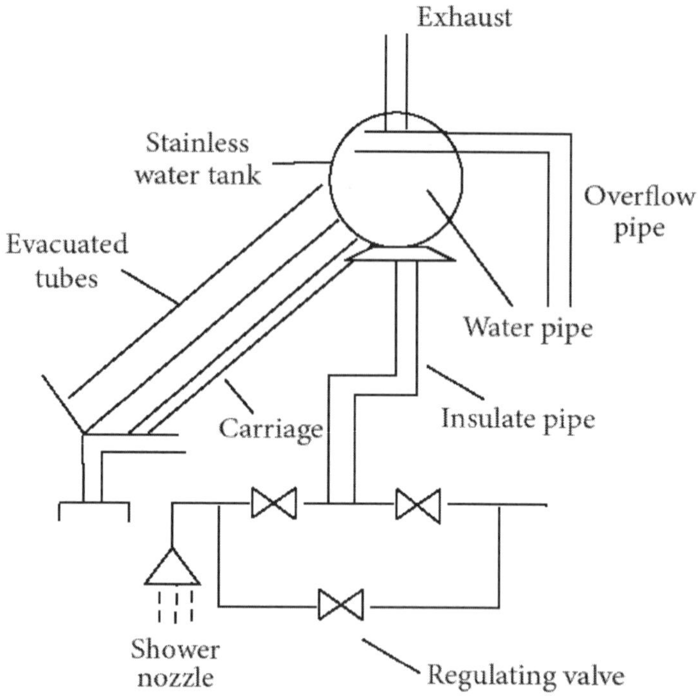

Figure 3: The constitutional diagram of solar heating water system.

3.3.2. The Selection of Solar Water Heating Systems for Service

Separation system is preferred from architectural aesthetic considerations; indirect system is preferred from water sanitation considerations; from water and energy saving considerations, the insulation measures must be taken into account according to the different forms of buildings.

Solar heating water system operation mode generally is determined by the user basic conditions, the relative needs of the user, the installation location of collector and storage tank, and other factors [16]. The recommended mode is shown in Table 2. The type of collector is based on the operation time of solar heating water system, the minimum temperature, and other factors.

By the comparison of flat plate collector and evacuated tubular collector, it can be seen that the economical efficiency, operation environment, useful life period, and stability of evacuated tube collector are all better than those of flat plate solar collector. Therefore all-glass evacuated tube collector is preferred for highway service. Solar water heating system should be installed in residential buildings, shower room, kitchen, and other building roofs or exterior for providing hot water for the service areas. The area of solar collector should also be designed based on the demand of hot water. When collector is installed on the roof, the designed tilt angle of collector is based on local solar elevation angle and the general criterion for the optimum tilt angle of solar collector is the local latitude plus 10 degrees [16].

Table 2: The comparison of recommended operating mode of solar heating water system for highway service area.

Operating conditions	Operating mode		
	Natural circulation	Direct circulation	Forced circulation
Instability water pressure	Available	Unavailable (1)	Available
Insufficient power supply	Available	Unavailable (2)	Unavailable (3)
Instant hot water	Unavailable	Available	Unavailable
The relative installation location of collector and storage tank			
High collector	Unavailable	Available	Available
High storage tank	Available	Available	Available
Operating temperature			
Above 0°C	Available	Available	Available
Below 0°C	Available with antifreezing measures		

(1) Available with thermostat control pump.
(2) Available with thermostatic valve.
(3) Available with DC pump under photoelectric cell.

4. RESULTS

4.1. The Application of Solar Photovoltaic Technology for Highway Service Area

4.1.1. Location and System Selection

The solar photovoltaic systems in West Lushan service were installed on the Northeast logistics building floor roof terrace, with the capacity of 20.24 kWp, medium-sized photovoltaic systems. In terms of the natural and geographical conditions and the demand for electricity of West Lushan service area, the independent DC with energy storage device solar photovoltaic power generation system was chosen as seen in Figures 5 and 6.

4.1.2. The Determination of Solar Panel Type and Installing Angle

The model of solar panel was S-220D (230 Wp), a total of 88 with a 220 V, 100 A solar controller as well as a 220 VDC/220 VAC 5 kVA separate inverter, which could meet the power supply for 3 consecutive rainy days as shown in Figure 4.

Figure 4: The solar panels of solar photovoltaic systems in West Lushan highway service area.

Figure 5: The solar controller of solar photovoltaic systems in West Lushan highway service area.

Figure 6: Solar photovoltaic power generation storage batteries of solar photovoltaic systems in West Lushan highway service area.

Considering the average amount of electricity use throughout the year, the tilt angle equalled the local latitude. The local latitude of West Lushan was 28.67 degrees, so the tilt angle was set as 28 degrees.

4.2. The Application of Solar Water Heating Technology for Highway Service Area

4.2.1. Selection of System Type and Installing Angle

Solar water heating systems were installed in the buildings of West Lushan highway service area for hand washing, showers, and other hot water needs like Lushan West Visitor Centre, dining lounge, logistics building, service stations, and maintenance station.

The direct system was chosen and the control mode was set with three modes of constant temperature control, difference temperature control, and timing control by freedom choice. The operation mode was forced circulation and the collector type was evacuated tubular collector based on the comparison analysis of flat plate collector and evacuated tubular collector for highway service area, as illustrated in Table 3.

Table 3: The comparison of flat plate collector and evacuated tubular collector for highway service area.

Features	Flat plate solar collector	Evacuated tubular solar collector
System efficiency	About 50% (summer) About 20% (winter)	About 50% (summer) About 40% (winter)
Useful life period	5 years	15 years
Operation temperature	Above 0°C	Above −40°C
Manufacturing cost	Low	Low
Key component efficiency	Fast heat absorption and fast heat release	Fast heat absorption and slow heat release
Efficiency in wet or rainy day	Low	High

According to the local solar elevation angle, the tilt angle of solar collector was the local latitude plus 10 degrees, which was 38 degrees.

4.2.2. The Solar Collector Installation Condition

The solar collector installation condition in West Lushan highway service area was shown in Table 4.

Table 4: List of installed solar collectors in West Lushan highway service area.

Installation location	Visitor centre	Dining lounge	Logistics building	Service station	Maintenance station	Total
Number of collectors	64	24	48	24	1	161
Size (each piece)			15 × Φ47 × 1500 mm			
Collecting area (m^2)	120	45	90	45	1.875	301.875

4.3. Economic Effects Analysis of Solar Energy Technology in West Lushan Highway Low-Carbon Service

4.3.1. The Power Generation Capacity of Solar Photovoltaic System

We can get the power generation capacity of solar photovoltaic system in West Lushan highway low-carbon service according to the following formula [17]:

$$G_{pv} = A_c \cdot J_r \cdot T \cdot n_b \cdot n_s, \tag{1}$$

where A_c is the effective area of photovoltaic cells m^2, J_r is the solar radiation intensity under the best tilt angle per year kw/m^2, T is the effective hours of sunshine per year h, is the component efficiency, and $n_s$is the system efficiency.

The parameters of solar water photovoltaic system in West Lushan service were as follows: the effective area of photovoltaic cells was about 140 m^2 in terms of the capacity of 20.24 kWp; the solar radiation intensity per year under the best tilt angle 28 degrees in West Lushan district was 1330 kw/m^2; The effective hours of sunshine per year was 1700.7; the component efficiency of silicon battery was generally set 15%; and the system efficiency was generally set 75%.

So the power generation capacity of solar photovoltaic system per year in West Lushan highway low-carbon service was 35625.41 kW·h.

4.3.2. Energy Saving Capacity of Solar Water Heating System

The total collecting area of solar water heating system in West Lushan highway service area was 301.875 m^2, which was calculated in terms of the following formula:

$$\Delta Q_{save} = A_c \cdot J_t \cdot (1 - \eta_c) \cdot \eta_{cd}, \quad (2)$$

where A_c is the solar collecting area of direct system m^2, J_t is the solar radiation of collecting surface per year MJ/m^2, n_c is the heat loss of pipeline and water tank, n_{cd} is the thermal efficiency in full daytime of solar collector %, and d is the number of days.

The parameters of solar water heating system in West Lushan service were as follows: the total solar collecting area in the service A_c was 301.875 m^2, solar radiation per year J_t was 4494.35 MJ/m^2, set n_c was 0.1, and set $n_{cd}$was 0.5.

So the amount of saving energy of solar water heating system in West Lushan service per year was 610530 MJ.

If the energy was generated by electricity heating, the consuming electricity was determined in terms of the following formula:

$$G = \frac{\Delta Q_{save}}{n_d q_0}, \tag{3}$$

where n_d is the efficiency of electricity heating and q_0 is the calorific value of electricity kJ/kW·h.

Generally, n_d is 0.9 and q_0 is 3595 kJ/kW·h.

So the electricity saving by the energy saving of solar water heating system per year is 188697.26 kW·h.

4.3.3. Economic Profit of Solar Energy Technology in West Lushan Highway Low-Carbon Service

Taking the investment of the solar water heating systems into account, the payback period of the investment of solar system was calculated according to the following formula:

$$T = \frac{y}{n_0}, \tag{4}$$

where Y is the investment of the system and $n_0$is the cost saving of the system.

According to the current electricity price of 0.60 RMB/kW·h in Jiangxi Province, the reduction of the cost by electricity savings of solar photovoltaic system in West Lushan highway service area was 21.38 thousand RMB per year. The cost of solar photovoltaic system per square meter was around 1300 RMB, so the total investment of solar photovoltaic system was 182 thousand RMB. Therefore, the payback period of the investment of solar photovoltaic system was 8.5 years. As we know the useful life period of solar photovoltaic system was at least 25 years, so economic profits of solar photovoltaic system were relatively high.

Taking 25 years as the effective operation period, the economic profit of the solar photovoltaic system in West Lushan highway service area would reach 352.5 thousand RMB.

The reduction of the cost by electricity savings of solar water heating system in West Lushan highway service area was 113.22 thousand RMB according to the current electricity price of 0.60 RMB/kW·h in Jiangxi Province. Because one piece of solar evacuated tubular collector was about 4000 RMB, the total investment of 161 pieces of solar evacuated tubular collector was 644 thousand RMB. Therefore, the payback period of the investment of solar water heating system was 5.69 years. As we know the useful life period of solar photovoltaic system was at least 15 years, so economic profits of solar photovoltaic system were very high.

Taking 15 years as the operation period, the economic profit of the solar photovoltaic system in West Lushan highway service area would reach 1.05 million RMB.

4.4. Environmental Benefits Analysis of Solar Energy Technology in West Lushan Highway Low-Carbon Service

The statistics from National Development and Reform Commission of China reveals that every saving of 1 kwh electricity equaled the consumption of 0.4 kg standard coal and also equaled a decrease of standard coal combustion producing 0.997 kg carbon dioxide (CO_2), 0.03 kg sulfur dioxide (SO_2), 0.015 kg nitrogen oxides (NO_x), 0.272 kg smoke, and other pollutants.

The energy saving of the solar technology in West Lushan highway service area equaled the reduction of 89.73 tons standard coal, also 233.65 tons CO_2, 6.73 tons SO_2, 3.36 tons NO_x, and 61.02 tons smoke. It could be seen that the solar technology of West Lushan highway service area would make great contributions to the reduction of environment pollution.

5. CONCLUSION

1) The closeness and independence of highway service area determined its dependence and pressure on resources and environment. The application of solar technology for highway service area could not only reduce energy consumption and improve energy efficiency, but also reduce air pollution and protect ecological environment.
2) Solar photovoltaic (PV) system was quite suitable for highway service area based on its advantages of small size and occupational

room, long useful life period, zero pollution emissions, and generation of power without water.

3) Based on the comparison analysis of solar photovoltaic systems and the natural and geographical conditions of West Lushan in Jiangxi Province, independent photovoltaic system with energy storage device was chosen for West Lushan highway service. And considering the average amount of electricity use throughout the year, the tilt angle of solar panel equalled the local latitude, set as 28 degrees according to the local latitude of West Lushan of 28.67 degrees.

4) It was proved that evacuated tubular collector and the operation mode of forced circulation were preferred for West Lushan highway service area by the comparison analysis. The selection of designed tilt angle of collector was based on local solar elevation angle and the tilt angle of solar collector in West Lushan highway service area was the local latitude plus 10 degrees, which was 38 degrees.

5) The power generation capacity of solar photovoltaic system per year in West Lushan highway low-carbon service was 35625.41 kW·h by calculation analysis. The reduction of the cost by electricity savings of solar photovoltaic system in West Lushan highway service area was 21.38 thousand RMB per year according to the current electricity price of 0.60 RMB/kW·h in Jiangxi Province.

Taking the investment of the solar water heating systems into account, the payback period of the investment of solar photovoltaic system was 8.5 years.

Taking 25 years as the effective operation period, the economic benefits of the solar photovoltaic system in West Lushan highway service area would reach 352.5 thousand RMB.

(6) The electricity saving by the saving energy of solar water heating system per year was 188697.26 kW·h. The reduction of the cost by electricity savings of solar water heating system in West Lushan highway service area was 113.22 thousand RMB per year.

Taking the investment into account, the payback period of the investment of solar water heating system was 5.69 years.

Taking 15 years as the operation period, the economic effect of the solar photovoltaic system in West Lushan highway service area would reach 1.05 million RMB.

(7) The energy saving of the solar technology in West Lushan highway service area equalled the reduction of 89.73 tons standard coal, also 233.65 tons CO_2, 6.73 tons SO_2, 3.36 tons NO_x, and 61.02 tons smoke. It could be seen that the solar technology of West Lushan highway service area would make great contributions to the reduction of environment pollution.

ACKNOWLEDGMENTS

The research work was supported by National Natural Science Foundation of China under Grant no. 51108216. The authors are very grateful for the helpful comments and criticisms of the anonymous reviewers.

REFERENCES

1. C. Shi, "The application of solar central heating systems in residential community of European cities,"Solar, vol. 4, pp. 18–20, 2003.
2. "Profession standard of Pepole's Repubilic of China," Technical Standards of Highway Construction JTG B01, 2003.
3. G. Hui and L. Liu, "The planning of Parking capacity and land size of rest facilities," Northeast Highway, 2006.
4. H. Luo, Study in Design of Highway Service, Northwest Highway, 2010.
5. R. Wu, "Case Study of wastewater treatment ecological technology for highway service," Environmental Science and Technology, vol. 11, pp. 96–98, 2007.
6. W. Yang, Study in wastewater biological contact oxidation treatment technology for highway service [M.S. thesis], Chang'an University, 2006.
7. A. D. Maria and D. Marano, "Tech-economic analysis of solar photovoltaic system," International Energy Journal, no. 5, pp. 47–68, 2004.
8. M. Schröder and B. Reddemann, "Three different criteria to evaluate the economics of solar water heating systems," Solar Energy, vol. 29, no. 6, pp. 549–555, 1982.
9. S.-S. Zilla and M. Abraham, "Simulation of a solar energy system for heating water," Computers and Industrial Engineering, vol. 8, pp. 1073–1084, 2004.

10. M. N. A. Hawlader, K. C. Ng, T. T. Chandratilleke, D. Sharma, and H. L. Kelvin Koay, "Economic evaluation of a solar water heating system," Energy Conversion and Management, vol. 27, no. 2, pp. 197–204, 1987.
11. J. Yi, Engergy Saving, Beijing China Architecture & Building Press, Beijing, China, 2006.
12. Ministry of Housing and Urban-Rural Development of the People's Republic of China. GBT 50378-2006 Green Building Assessment Standard. Beijing China Architecture & Building Press, 2006.
13. M.-S. Hsu and F.-J. Lin, "The developing strategy of green energy industry cluster A case study of the solar photoelectric industry in Taiwan," Procedia—Social and Behavioral Sciences, vol. 40, pp. 165–173, 2012.
14. V. V. Tyagi, N. A. A. Rahim, and J. A. L. Selvaraj, "Progress in solar PV technology: research and achievement," Renewable and Sustainable Energy Reviews, vol. 20, pp. 443–461, 2013.
15. C. L. Bensona and C. L. Mageea, "On improvement rates for renewable energy technologies solar PV, wind turbines, capacitors," Renewable Energy, vol. 68, pp. 745–751, 2014.
16. Ministry of Housing and Urban-Rural Development of the People's Republic of China, GBT 50378-2006 Civil Architecture Soar Water Heating System Application Standard, Architecture & Building Press, Beijing, China, 2005.
17. J. Wang, Study on the economic evaluation of solar water heating system for building in Wuhan [Master's, thesis], Wuhang University, Wuhan, China, 2012.

12
CHAPTER

SOLAR CELLS: IN RESEARCH AND APPLICATIONS—A REVIEW

Shruti Sharma[1], Kamlesh Kumar Jain[1], Ashutosh Sharma[2*]

[1]*CMD College, Bilaspur, India*

[2]*Materials Science and Engineering, University of Seoul, Seoul, South Korea*

ABSTRACT

The light from the Sun is a non-vanishing renewable source of energy which is free from environmental pollution and noise. It can easily compensate the energy drawn from the non-renewable sources of energy such as fossil fuels and petroleum deposits inside the earth. The fabrication of solar cells has passed through a large number of improvement steps from one generation to another. Silicon based solar cells were the first generation solar cells grown on Si wafers, mainly single crystals. Further development to thin films, dye sensitized solar cells and organic solar cells enhanced the cell efficiency. The development is basically hindered by the cost and efficiency. In order to choose the right solar cell for a specific geographic location, we are required to understand fundamental mechanisms and functions of several solar technologies that are widely studied. In this article, we have reviewed a progressive development in the solar cell research from one generation to other, and discussed about their future trends and aspects. The article also tries to emphasize the various practices and methods to promote the benefits of solar energy.

Keywords: Solar Energy, Fossils, Power, Pollution, Renewable

1. INTRODUCTION

Everyday sun sends out tremendous amount of energy in the form of heat and radiations called solar energy. Solar energy is a limitless source of energy which is available at no cost [1] [2] . The major benefit of solar energy over other conventional power generators is that the sunlight can be directly harvested into solar energy with the use of small and tiny photovoltaic (PV) solar cells [3] [4] . The Sun is assumed as a big spherical gaseous cloud made up of hydrogen and helium atoms. This big spherical gaseous cloud is mainly composed of several hydrogen nu lei combining to form helium energy with the emission of energy from the fusion of the hydrogen nuclei in inner core of the Sun via nuclear fusion (Figure 1) [5] .

$$ {}_{1}^{2}H + {}_{1}^{2}H \rightarrow {}_{2}^{3}He + {}_{0}^{1}n + 3.27\ \text{MeV} \tag{1} $$

$$ {}_{1}^{2}H + {}_{1}^{3}H \rightarrow {}_{2}^{4}He + {}_{0}^{1}n + 17.59\ \text{MeV} \tag{2} $$

During this process of fusion, four hydrogen atoms combine to form one helium atom with a loss of mass which is radiated as thermal energy [1] [4] - [5] . This radiant energy produced by fusion reactions is free from any pollutant, gases, or other reaction by-product. This is why it is the major driving force of all the clean energy technology, in view of the climatic disturbance caused by the emission of carbon from the fossil fuels deposits. One of the biggest advantages of solar energy is that it is free reachable to common people and available in abundant supply compared to that of the price of various fossil fuels and oils in the past decade [1] - [7] . Moreover, solar energy requires considerably lower manpower expenses over conventional energy production technology.

Though the solar energy is freely available everywhere, there is still an initial expenditure on the equipments for harvesting this radiant energy by developing solar cells, panels and modules [8] . These small and tiny solar cells produce no noise during their operation. On the other hand, the big power pumping devices produce unbearable sound pollution, and therefore they are very disturbing to the society [6] -[8] . Nowadays, due to the decreasing amount of renewable energy resources, the per watt cost of solar energy device has become more important in the last decade, and is definitely set to become economical in the coming years and grow as better technology in terms of both cost and applications [9] [10] .

In spite of numerous advantages, this energy has few limitations too. Firstly, solar energy doesn't radiate at night. Secondly, the solar energy is almost not constant all the time. There must be plenty of sunlight available to generate electrical energy from a solar PV device [7] [10] [11] . Moreover, apart from daily fluctuations in the intensity of radiant energy, the solar energy is hindered to reach the earth during bad climatic conditions. For example, the amount of sunlight reaching the earth's surface depends on location, time as well as weather as it falls during winter season as compared to the summer, and the Sun's radiation is less intense [10] [11] . To overcome these demerits of this technology, solar energy must be stored elsewhere at night and the highly efficient solar cells and modules needs to be developed.

There have been an enormous amount of research activities to harvest the Sun's energy effectively by developing solar cells/panels with high conversion efficiencies. The photovoltaic conversion efficiency is referred to the efficiency of solar PV modules, and is defined as the fraction of Sun's energy that can be converted into electricity. Solar panels are a huge collection of tiny solar cells arranged in a definite geometrical shape to produce a given amount of power supply. The storage of solar power is still has not been achieved successfully. Currently the radiation efficiency of solar panel is up to 22% [11] . There are many solar photovoltaic batteries available which are usually more expensive and bulky. These are more suitable for small scale or household solar needs compared to large solar plants [12] [13] .

The working mechanism of solar cells is based on the three factors: (1) Adsorption of light in order to generate the charge careers, holes (p-type) and electrons (n-type) (2) Separation of charge careers, and (2) the collection of charge careers at the respective electrodes establishing the potential difference across the p-n junction. The generation of voltage difference noticed at the p-n junction of the cell in response to visible radiation is utilized to do the work.

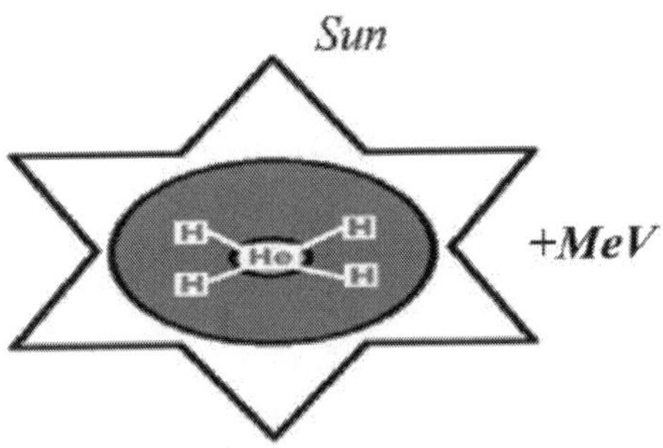

$$^{2}_{1}H + ^{2}_{1}H \rightarrow ^{3}_{2}He + ^{1}_{0}n + 3.27\text{MeV}$$

$$^{2}_{1}H + ^{3}_{1}H \rightarrow ^{4}_{2}He + ^{1}_{0}n + 17.59\text{MeV}$$

Figure 1. Nuclear fusion reaction: source of solar energy [5] .

In the past, various kinds of semiconductor materials and technologies are devised to design solar cells with low cost as well as high conversion efficiency. Traditional solar panels made from silicon crystalline wafer modules are heavier which makes the transportation difficult. These are generally the large sized solar panels covered with glass sheets. A heavier and bulky solar panel requires a lot of space and sometime big roofs to fit these bulky and large solar panels in case of high power applications [11] - [13] . Therefore, in this article keeping in mind the efficient use of solar energy by solar cells research and development, we will study the different types of solar cells.

2. SOLAR CELLS

The photovoltaic (PV) effect was first observed by Alexandre-Edmond Becquerel in 1839 [14] . Subsequently, in 1946 the first modern solar cell made of silicon was invented by Russel Ohl [14] [15] . Earlier photovoltaic solar cells are thin silicon wafers that transform sunlight energy into electrical power. The modern photovoltaic technology is based on the principle of electron hole creation in each cell composed of two different layers (p-type and n-type materials) of a semiconductor material, as shown in Figure 2. In this arrangement of the structure, when a photon of sufficient energy impinges on the p-type and n-type junction, an electron is ejected by gaining energy from the striking photon and moves from one layer to another. This creates an electron and a hole in the process and by this process electrical power is generated [16] . The various types of materials applied for photovoltaic solar cells includes mainly in the form of silicon (single crystal, multi-crystalline, amorphous silicon) [3] - [6] , cadmium-telluride [3] [4] , copper-indium-gallium-selenide [3] [4] [8] , and copper-indium-gallium-sul- fide [10] [16] . On the basis of these materials, the photovoltaic solar cells are categorized into various classes as discussed in the following sections (also shown in Figure 3).

2.1. First Generation Solar Cell—Wafer Based

As it is already mentioned, the first generation solar cells are produced on silicon wafers. It is the oldest and the most popular technology due to high power efficiencies. The silicon wafer based technology is further categorized into two subgroups named as [2] - [4] [10] [16] .

- Single/ Mono-crystalline silicon solar cell.
- Poly/Multi-crystalline silicon solar cell.

2.1.1. Single/Mono-Crystalline Silicon Solar Cell

Mono crystalline solar cell, as the name indicates, is manufactured from single crystals of silicon by a process called Czochralski process [16] - [18] . During the manufacturing process, Si crystals are sliced from the big sized ingots. These large single crystal productions require precise processing as the process of "recrystallizing" the cell is more expensive and multi process. The efficiency of mono-crystalline single-crystalline silicon solar cells lies between 17% - 18% [8] .

2.1.2. Polycrystalline Silicon Solar Cell (Poly-Si or Mc-Si)

Polycrystalline PV modules are generally composed of a number of different crystals, coupled to one another in a single cell. The processing of polycrystalline Si solar cells is more economical, which are produced by cooling a graphite mold filled containing molten silicon. Polycrystalline Si solar cells are currently the most popular solar cells. They are believed to occupy most up to 48% of the solar cell production worldwide during 2008 [19] . During solidification of the molten silicon, various crystal structures are formed. Though they are slightly cheaper to fabricate compared to monocrystalline silicon solar panels, yet are less efficient ~12% - 14% [20] .

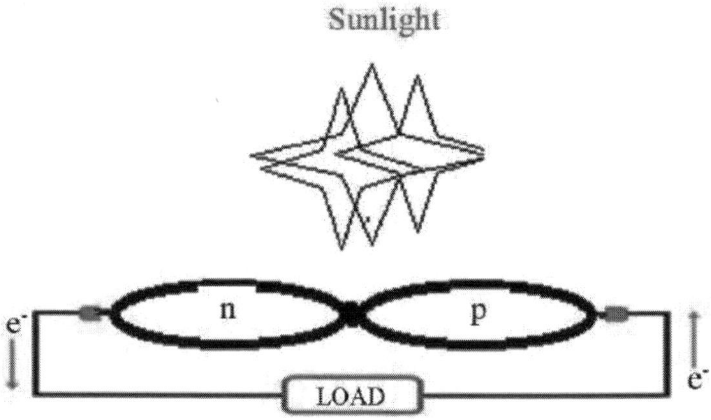

Figure 2. The semiconductor p-n junction solar cell under load.

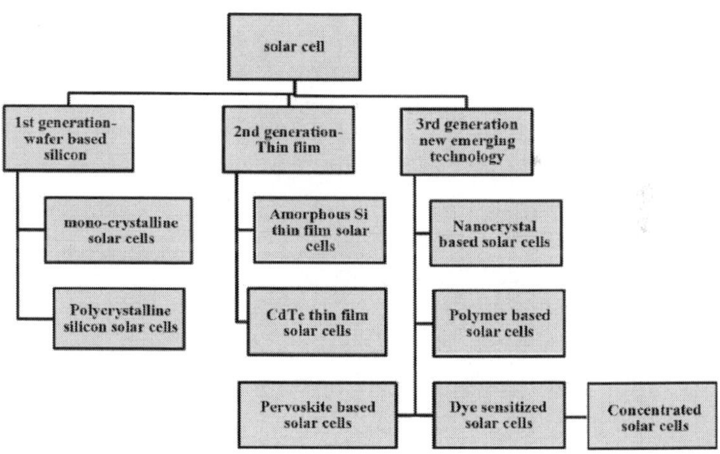

Figure 3. Various types of solar cell technologies and current trends of development [2] [16] .

2.2. Second Generation Solar Cells—Thin Film Solar Cells

Most of the thin film solar cells and a-Si are second generation solar cells, and are more economical as compared to the first generation silicon wafer solar cells. Silicon-wafer cells have light absorbing layers up to 350 μm thick, while thin-film solar cells have a very thin light absorbing layers, generally of the order of 1 μm thickness [21] . Thin film solar cells are classified as;

- a-Si.
- CdTe.
- CIGS (copper indium gallium di-selenide).

2.2.1 Amorphous Silicon Thin Film (a-Si) Solar Cell

Amorphous Si (a-Si) PV modules are the primitive solar cells that are first to be manufactured industrially. Amorphous (a-Si) solar cells can be manufactured at a low processing temperature, thereby permitting the use of various low cost, polymer and other flexible substrates. These substrates require a smaller amount of energy for processing [22] . Therefore, a-Si amorphous solar cell is comparatively cheaper and widely available. The "amorphous" word with respect to solar cell means that the comprising silicon material of the cell lacks a definite arrangement of atoms in the lattice, non-crystalline structure, or not highly structured. These are fabricated by coating the doped silicon material to the backside of the substrate/glass plate. These solar cells generally are dark brown in color on the reflecting side while silverish on the conducting side [23] .

The mainissue of a-Si solar cell is the poor and almost unstable efficiency. The cell efficiency automatically falls at PV module level. Currently, the efficiencies of commercial PV modules vary in the range of 4% - 8%. They can be easily operated at elevated temperatures, and are suitable for the changing climatic conditions where sun shines for few hours [24] .

2.2.2 Cadmium Telluride (CdTe) Thin Film Solar Cell

Among thin-film solar cells, cadmium telluride (CdTe) is one of the leading candidate for the development of cheaper, economically viable photovoltaic (PV) devices, and it is also the first PV technology at a low cost [8] [25] [26] . CdTe has a band gap of ~1.5 eV as well as high optical absorption coefficient and chemical stability. These properties make CdTe most attractive material for designing of thin-film solar cells.

CdTe is an excellent direct band gap crystalline compound semiconductor which makes the absorption of light easier and improves the efficiency. It is generally constructed by sandwiching between cadmium sulfide layers to form a p-n junction diode. The manufacturing process involves three steps: Firstly, the CdTe based solar cells are synthesized from polycrystalline materials and glass is chosen a substrate. Second process involves deposition, i.e., the multiple layers of CdTe solar cells are coatedon to substrate using different economical methods. It is already mentioned that CdTe has a direct optimum band gap (~1.45 eV) with high absorption coefficient over 5×10^{15}/cm [27] . Therefore, its efficiency usually operates in the range 9% - 11% [8] [28] . CdTe solar cells can be made on polymer substrates and flexible. However, there are various environmental issues with cadmium component of solar cell. Cadmium is regarded as a heavy metal and potential toxic agent that can accumulate in human bodies, animals and plants. The disposal of the toxic Cd based materials as well as their recycling can be highly expensive and damaging too to our environment and society [10] [24] . Therefore, a limited supply of cadmium and environmental hazard associated with its use are the main issues with this CdTe technology [26] - [29] .

2.2.3 Copper Indium Gallium Di-Selenide (CIGS) Solar Cells

CIGS is a quaternary compound semiconductor comprising of the four elements, namely: Copper, Indium, Gallium and Selenium [10] [30] . CIGS are also direct band gap type semiconductors. Compared to the CdTe thin film solar cell, CIGS hold a higher efficiency ~10% - 12%. Due to their significantly high efficiency and economy, CIGS based solar cell technology forms one of the most likely thin film technologies. The processing of CIGS are done by the following techniques: sputtering, evaporation, electrochemical coating technique, printing and electron beam deposition [16] [31] . In addition, the sputtering can be a two or multi-step process involving with deposition and subsequent interaction with selenium later, or can be a one-step reactive process. However, evaporation is similar to the sputtering in the sense that it can be used in a single step, two-step or multiple processing steps. The substrates for CIGS material can be chosen from glass plate, polymers substrates, steel, aluminum etc. The advantages of CIGS thin film solar cells include its prolonged life without a considerable degradation. These properties of CIGS indicate an easy solution to enhance the efficiency [8] [22] [28] . Table 1 shows a comparison of popular wafer based solar cell and thin film solar cell [16] [17] .

2.3. Third Generation Solar Cells

Third generation cells are the new promising technologies but are not commercially investigated in detail. Most of the developed 3rd generation solar cell types are [2] :

1) Nano crystal based solar cells.
2) Polymer based solar cells.
3) Dye sensitized solar cells.
4) Concentrated solar cells.

2.3.1. Nano Crystal Based Solar Cells

Nanocrystal based solar cells are generally also known as Quantum dots (QD) solar cells. These solar cells are composed of a semiconductor, generally from transition metal groups which are in the size of nanocrystal range made of semiconducting materials. QD is just a name of the crystal size ranging typically within a few nanometers in size, for example, materials like porous Si or porous TiO_2, which are frequently used in QD [32] . The structure of the QD solar cells are shown in Figure 4 [10] . With the advance of nanotechnology, these nanocrystals of semiconducting material are targeted to replace the semiconducting material in bulk state such as Si, CdTe or CIGS. This idea of the QD based solar cell with a theoretical formulation were employed for the design of a p-i-n solar cell over the self-organized in As/GaAs system [32] . Generally, the nanocrystals are mixed into a bath and coated onto the Si substrate. These crystals rotate very fast and flow away due to the centrifugal force. In conventional compound semiconductor solar cells, generally a photon will excite an electron there by creating one electron-hole pair [33] . However, when a photon strikes a QD made of the similar semiconductor material, numerous electron-hole pairs can be formed, usually 2 or 3, also 7 has been observed in few cases [2] [29] .

2.3.2. Polymer Solar Cells

Polymer solar cells (PSC) are generally flexible solar cells due to the polymer substrate. The first PSC were invented by the research group of Tang et al. at Kodak Research Lab. [2] . A PSC is composed of a serially connected thin functional layers coated on a polymer foil or ribbon. It works usually as a combination of donor (polymer) and a acceptor (fullerene). There are various types of materials for the absorption of sunlight, including organic material like a conjugate/conducting polyme [2] [34] . In 2000, Heeger, MacDiarmid, and Shirakawa fetched the Nobel Prize in Chemistry for the discovering a new category of polymer materials known as conducting polymers [35] [36] . The PSC and other organic solar cells operate on same principle known as the photovoltaic effect, i.e., where the transformation of the energy occurs in the form of electromagnetic radiations into electrical current [37] . Yu et al. mixedpoly [2-methoxy-5-(2'-ethylhexyloxy)-p-phenylene vinylene] (PPV), C60 and its other derivatives to develop the first polymer solar cell and obtained a high power conversion efficiency [38] . This process triggered the development of a new age in the polymer materials for capturing the solar power. After significantly optimizing the parameters, researchers achieved efficiency over 3.0% for PPV type PSCs [38] - [40] . These unique properties of PSCs opened a new gateway for new applications in the formation of stretchable solar devices including textiles and fabrics [39] . A modern recycling concept known as polarizing organic photovoltaics (ZOPVs) was also developed for increasing the function of liquid crystal displays utilizing the same polarizer, a photovoltaic device and proper light conditions/solar panel [39] - [41] .

Table 1.A comparison of various types of aolar cells[16][17].

Cell type	Crystalline silicon		Thin Film			Third Generation				Perovskites
	Monocrystalline	Polycrystalline	CdTe	CIGS	Amorphous Silicon	Nanocrystal	Dye Sensitized	Polymer	Concentrated	
Efficiency	14% - 17.5%	12% - 14%	9% - 11%	10% - 12%	4% - 8%	7% - 8%	~10%	~3% - 10%	~40%	31%
High temperature performance	Not good at high temperatures	Not good at high temperatures	Good in cool as well as high temperature conditions	Good in cool as well as high temperature conditions	Good in cool as well as high temperature conditions	Excellent thermal stability	Not good in high temperature conditions	Not good in high temperature conditions	Excellent thermal stability	Excellent thermal stability
Size	Significantly less volume to produce the same amount of power	Significantly less volume to produce the same amount of power	Offering a wide range of product design from flexible, light durable	Offering a wide range of product design from flexible, light durable	Offering a wide range of product design from flexible, light durable	Offering a wide range of product design from flexible, light durable	Offering a wide range of product design from flexible, light durable	Offering a wide range of product design from flexible, light durable	Offering a specialized range of product design	Offering a wide range of product design from flexible, light durable
Cost	Two times more expensive compared to thin-film.	Two times more expensive compared to thin-film.	50 percent less expensive than conventional silicon cells	50 percent less expensive than conventional silicon cells	50 percent less expensive than conventional silicon cells	50 percent less expensive than conventional silicon cells	50 percent less expensive than conventional silicon cells	50 percent less expensive than conventional silicon cells	50 percent less expensive than conventional silicon cells	50 percent less Expensive than conventional silicon cells
Additional detail	Oldest PV technology	Economical choice	Toxic due to Cd	Some CIGS have impressive 20% efficiency	Needs long installacion time and large space	Needs short installation time and large space	Needs short installation time and large space	Needs short installation time and small space	Needs long installation time and large space	Latest technology. Needs short installation time and minimum space

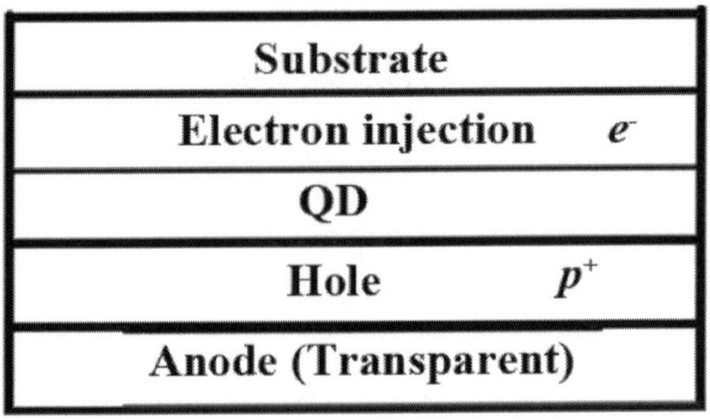

Figure 4. A schematic of Quantum dot (QD) layer [32] .

2.3.3. Dye Sensitized Solar Cells (DSSC)

Recent research has been focused on improving solar efficiency by molecular manipulation, use of nanotechnology for harvesting light energy [42] - [45] . The first DSSC solar cell was introduced by Michel Gratzel in Swiss federal institute of technology [10] [16] . DSSCs based solar cells generally employ dye molecules between the different electrodes. The DSSC device consists of four components: semiconductor electrode (n-type TiO_2 and p-type NiO), a dye sensitizer, redox mediator, and a counter electrode (carbon or Pt) [46] . The DSSCs attractive due to the simple conventional processing methods like printing techniques, are highly flexible, transparent and low cost as well [10] . The novelty in the DSSC solar cells arise due to the photosensitization of nano grained TiO_2 coatings coupled with the visible optically active dyes, thus increasing the efficiencies greater than 10% [42] - [44] [46] [47] . However, there are certain challenges like degradation of dye molecules and hence stability issues [10] . This is due to poor optical absorption of sensitizers which results in poor conversion efficiency. The dye molecules generally degrade after exposure to ultraviolet and infrared radiations leading to a decrease in the lifetime and stability of the cells. Moreover, coating with a barrier layer may also increase the manufacturing more expensive and lower the efficiency [8] .

2.3.4. Concentrated Solar Cells

Concentrating photovoltaic (CPV) has been established since the 1970s [45] [48] [49] . It is the newest technology in the solar cell research and development. The main principle of concentrated cells is to collect a large amount of solar energy onto a tiny region over the PV solar cell, as shown in Figure 5. The principle of this technology is based on optics, by using large mirrors and lens arrangement to focus sunlight rays onto a small region on the solar cell [8] . The converging of the sunlight radiations thus produces a large amount of heat energy. This heat energy is further driven by a heat engine controlled by a power generator with integrated. CPVs have shown their promising nature in solar world [49] [50] . It ca n be classified into low, medium, and high concentrated solar cells depending on the power of the lens systems [49] . Concentrating photovoltaic technology have the following merits, such as solar cell efficiencies >40%, absence of any moving parts, no thermal mass, speedy response time and can be scalable to a range of sizes.

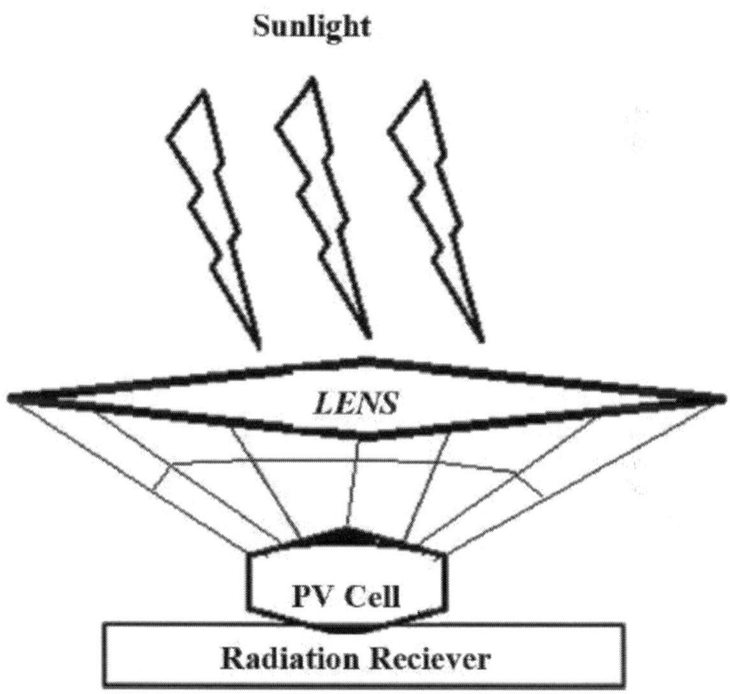

Figure 5. Schematic of concentrated solar cell [48] [49] .

2.4. Perovskite Based Solar Cell

Perovskites are a class of compounds defined by the formula ABX_3 where X represents a halogen such as I^-, Br^-, Cl^-. and A and B are cations of different size. Perovskite solar cells are recent discovery among the solar cell research community and possess several advantages over conventional silicon and thin film based solar cells. Conventional Si based solar cells need expensive, multiple processing steps and require high temperatures (>1000°C) and vacuums facilities [51] [52] . The perovskites based solar cells can have efficiency up to 31% [53] . It can be predicted that these perovskites may also play an important role in next-generation electric automobiles batteries, according to an interesting investigation recently performed by Volkswagen [52] [53] . However, current issues with perovskite solar cells are their stability and durability. The material degrades over time, and hence a drop in overall efficiency. Therefore more research is needed to bring these cells into the market place.

3. ADVANCES IN ENERGY STORAGE

Since the sunlight is not always available, all these businesses of PV solar cells may not work at night and a lot of electricity will go unused [54] . Therefore energy storage is an important factor in solar cell market. A comparison and summary of various types of solar cells is summarized in Table 1.

Several energy storage devices are available in the market but those are highly expensive and a short life span. Recently, in 2014, Harvard University researchers developed a new type of battery based on organic molecules called Quinone. It is found in plants and is economical in a sense that it can store sunlight energy for a couple of days [55] .

The world's first solar cell energy storage is introduced by Wu and his co-workers at Ohio State University. This device not only can store energy but can also reduce the costs of renewable energy by 25%, relying on a new aqueous, rechargeable lithium-oxygen battery used in sunlight [56] .

4. CONCLUSION

Solar power generation has been developed as one of the most demanding renewable sources of electricity. It has several advantages compared to other forms of energy like fossils fuels and petroleum deposits. It is an alternative which is promising and consistent to meet the high energy demand. Though the methods of utilizing solar energy are simple, yet need an efficient and durable solar material. Technology based on nano-crystal QD of semiconductors based solar cell can theoretically convert more than sixty percent of the whole solar spectrum into electric power. The polymer base solar cells are also a viable option. However, their degradation over time is a serious concern. There are various challenges for this industry, including lowering the cost of production, public awareness and best infrastructure. Solar energy is the need of the day and research on the solar cells has a promising future worldwide.

REFERENCES

1. Chu, Y. and Meisen, P. (2011) Review and Comparison of Different Solar Energy Technologies. Report of Global Energy Network Institute (GENI), Diego.
2. Choubey, P.C., Oudhia, A. and Dewangan, R. (2012) A Review: Solar Cell Current Scenario and Future Trends. Recent Research in Science and Technology, 4, 99-101.
3. McEvoy, A., Castaner, L. and Markvart, T. (2012) Solar Cells: Materials, Manufacture and Operation. 2nd Edition, Elsevier Ltd., Oxford, 3-25.
4. Fahrenbruch, A.L. and Bube, R.H. (1983) Fundamentals of Solar Cells. Academic Press Inc., New York.
5. (2015-2016) Energy from the Sun, Student Guide. National Energy Education Development Project (NEED).
6. Grisham, L.R. (2008) Nuclear Fusion in: Future Energy, Improved, Sustainable and Clean Options for our Planet, Edited by Trevor M. Letcher, 2nd Edition, Elsevier Ltd., Amsterdam, 291-301.
7. Rana, S. (2013) A Study on Automatic Dual Axis Solar Tracker System using 555 Timer. International Journal of Scientific & Technology Research, 1, 77-85.
8. Bertolli, M. (2008) Solar Cell Materials. Course: Solid State II. Department of Physics, University of Tennessee, Knoxville.
9. Wall, A. (2014) Advantages and Disadvantages of Solar Energy. Process Industry Forum, 7 August 2013. Web. 2 February 2014.
10. Bagher, A.M., Vahid, M.M.A. and Mohsen, M. (2015) Types of Solar Cells and Application. American Journal of Optics and Photonics, 3, 94-113.

11. Whitburn, G. (2012) Exploring Green Technology, Fundamental Advantages and Disadvantages of Solar Energy.
12. Peplow, M. (2014) Organic Synthesis: The Robo-Chemist. Nature, 512, 20-22.
13. Tsoutsos, T., Frantzeskaki, N. and Gekas, V. (2005) Environmental Impacts from the Solar Energy Technologies. Energy Policy, 33, 289-296.
14. Yadav, A. and Kumar, P. (2015) Enhancement in Efficiency of PV Cell through P&O Algorithm. International Journal for Technological Research in Engineering, 2, 2642-2644.
15. Castellano, R. (2010) Solar Panel Processing. Old City Publishing Inc., Philadelphia.
16. Srinivas, B., Balaji, S., Nagendra Babu, M. and Reddy, Y.S. (2015) Review on Present and Advance Materials for Solar Cells. International Journal of Engineering Research-Online, 3, 178-182.
17. Wurfel, P. and Wurfel, U. (2009) Physics of Solar Cells: From Basic Principles to Advanced Concepts. John Wiley & Sons, Hoboken.
18. Dmitrijev, S. (2006) Principles of Semiconductor Devices. Oxford University Press, Oxford.
19. Saga, T. (2010) Advances in Crystalline Silicon Solar Cell Technology for Industrial Mass Production. NPG Asia Materials, 2, 96-102.
20. Jayakumar, P. (2009) Solar Energy Resource Assessment Handbook. Renewable Energy Corporation Network for the Asia Pacific.
21. Chopra, K.L., Paulson, P.D. and Dutt, V. (2004) Thin-Film Solar Cells: An Overview. Progress in Photovoltaics, 12, 69-92.
22. Imamzai, M., Aghaei, M., Hanum Md Thayoob, Y. and Forouzanfar, M. (2012) A Review on Comparison between Traditional Silicon Solar Cells and Thin-Film CdTe Solar Cells. Proceedings of National Graduate Conference (NatGrad 2012), Tenaga Nasional Universiti, Putrajaya Campus, 8-10 November 2012, 1-5.
23. Types of Solar Panels, Grein Energy. Published 22 April 2015.
24. Maehlum, M.A. (2015) Energy Informative The Homeowner's Guide To Solar Panels, Best Thin Film Solar Panels— Amorphous, Cadmium Telluride or CIGS? Last updated 6 April 2015.
25. Yogi Goswami, D. and Kreith, F. (2007) Handbook of Energy Efficiency and Renewable Energy. CRC Press, Boca Raton.

26. Luque, A. and Hegedus, S. (2003) Handbook of Photovoltaic Science and Engineering. 2nd Edition, John Wiley & Sons, Ltd., Hoboken.
27. Elsabawy, K.M., El-Hawary, W.F. and Refat, M.S. (2012) Advanced Synthesis of Titanium-Doped-Tellerium-Camium Mixtures for High Performance Solar Cell Applications as One of Renewable Source of Energy. International Journal of Chemical Sciences, 10, 1869-1879.
28. Badawy, W.A. (2015) A Review on Solar Cells from Si-Single Crystals to Porous Materials and Quantum Dots. Journal of Advanced Research, 6, 123-132.
29. Sethi, V.K., Pandey, M. and Shukla, P. (2011) Use of Nanotechnology in Solar PV Cell. International Journal of Chemical Engineering and Applications, 2, 77-80.
30. Andorka, F. (2014) CIGS Solar Cells Simplified. Solar Power World. http://www.solarpowerworldonline.com/2014/01/cigs-solar-cells-simplified/
31. Razykov, T.M., Ferekides, C.S., Morel, D., Stefanakos, E., Ullal, H.S. and Upadhyaya, H.M. (2011) Solar Photovoltaic Electricity: Current Status and Future Prospects. Solar Energy, 85, 1580-1608.
32. Hoppe, H. and Sariciftci, N.S. (2008) Polymer Solar Cells. Advances in Polymer Science, 214, 1.
33. Dubey, S., Sarvaiya, J.N. and Seshadri, B. (2013) Temperature Dependent Photovoltaic (PV) Efficiency and Its Effect on PV Production in the World: A Review. Energy Procedia, 33, 311-321.
34. Ganesh, B.N.V.S. and Supriya, Y.V. (2013) Recent Advancements and Techniques in Manufacture of Solar Cells: Organic Solar Cells. International Journal of Electronics and Computer Science Engineering, 2, 565-573.
35. Jassen, R. Introduction to Polymer Solar Cells. http://www.eet.bme.hu/~plesz/Gyarmati%20Adam/2.pdf
36. Leroy, F. (2003) A Century of Nobel Prize Recipients: Chemistry, Physics, and Medicine. CRC Press, Hoboken.
37. Alex, C., Mayer, Shawn, R., Scully, Brian, E., Hardin, Michael, W., Rowell and Michael, D. McGehee (2007) Polymer-Based Solar Cells: A Review. Materials Today, 10, 28-33.
38. Wudl, F. and Srdanov, G. (1993) Conducting Polymer Formed of Poly (2-Methoxy-5-(2'-Ethylhexyloxy)-P-Phenylene Vinylene). US Patent 5,189,136.

39. Li, G., Zhu, R. and Yang, Y. (2012) Polymer Solar Cells. Nature Photonics, 6, 153-161.
40. Brabec, C.J., Shaheen, S.E., Winder, C. and Sariciftci, N.S. (2002) Effect of LiF/Metal Electrodes on the Performance of Plastic Solar Cells. Applied Physics Letters, 80, 1288.
41. Zhu, R., Kumar, A. and Yang, Y. (2011) Polarizing Organic Photovoltaics. Advanced Materials, 23, 4193-4198.
42. Li, B., Wang, L., Kang, B., Wang, P. and Qiu, Y. (2006) Review of Recent Progress in Solid-State Dye-Sensitized Solar Cells. Solar Energy Materials and Solar Cells, 90, 549-573.
43. Zhan, J., Sun, P., Jiang, S., Sun, X. and Lund, T. (2006) An Investigation of the Performance of Dye-Sensitized Nanocrytsalline Solar Cell with Anthocyanin Dye and Ruthenium Dye as the Sensitizers. Roskilde University, Roskilde.
44. Graetzel, M., Janssen, R.A.J., Mitzi, D.B. and Sargent, E.H. (2012) Materials Interface Engineering for Solution-Processed Photovoltaics. Nature, 488, 304-312.
45. Nozik, A.J. (2010) Nanoscience and Nanostructures for Photovoltaics and Solar Fuels. NANO Letters, 10, 2735-2741.
46. Suhaimi, S., Shahimin, M.M., Alahmed, Z.A., Chysky, J. and Reshak, A.H. (2015) Materials for Enhanced Dye-Sensitized Solar Cell Performance: Electrochemical Application. International Journal of Electrochemical Science, 10, 28-59.
47. Liang, M., Xu, W., Cai, F.S., Chen, P.Q., Peng, B., Chen, J. and Li, Z.M. (2007) New Triphenylamine-Based Organic Dyes for Efficient Dye-Sensitized Solar Cells. The Journal of Physical Chemistry C, 111, 4465-4472.
48. Barnes, P.R.F., Anderson, A.Y., Koops, S.E., Durrant, J.R. and O'Regan, B.C. (2009) Electron Injection Efficiency and Diffusion Length in Dyesensitized Solar Cells Derived from Incident Photon Conversion Efficiency Measurements. The Journal of Physical Chemistry C, 113, 1126-1136.
49. Philipps, S.P., Bett, A.W., Horowitz, K. and Kurtz, S. (2015) Current Status of Concentrator Photovoltaics (CPV) Technology. Report Version 1.2, Fraunhofer Institute for Solar Energy Systems (NREL), September 2015.

50. Mohanta, P. R., Patel, J., Bhuva, J. and Gandhi, M. (2015) A Review on Solar Photovoltaics and Roof Top Application of It. International Journal of Advanced Research in Science, Engineering and Technology, 2, 2394-2444.
51. Ahn, N., Son, D.-Y., Jang, I.-H., Kang, S.M., Choi, M. and Park, N.-G. (2015) Highly Reproducible Perovskite Solar Cells with Average Efficiency of 18.3% and Best Efficiency of 19.7% Fabricated via Lewis Base Adduct of Lead(II) Iodide. Journal of the American Chemical Society, 137, 8696-8699.
52. Tina Casey (2015) An Article on Perovskites Will Power New Low-Cost & Highly Efficient Solar Cells. Clean Technical, 3 July 2015.
53. Shi, D., Zeng, Y. and Shen, W. (2015) Pervoskite/c-Si Tandem Solar Cell with Inverted Nanopyramids: Realizing High Efficiency by Controllable Light Trapping. Scientific Reports, 5, Article No. 16504.
54. http://www.altenergy.org/renewables/solar/latest-solar-technology.html
55. Wu, Y. and Gorder, P.F. (2014) Nature Communications. Published on 3 October 2014.
56. Nabhani, N. and Emami, M. (2013) Nanotechnologies and Its Applications in Solar Cells. International Conference on Mechanical and Industrial Engineering (ICMIE'2013), Penang, 28-29 August 2013, 88-91.

INDEX